S0-BDM-927

ENERGY AND CLIMATE CHANGE

Report of the DOE Multi-Laboratory Climate Change Committee

Lawrence Livermore National Laboratory
Chair
Argonne National Laboratory
Battelle Pacific Northwest Laboratory
Brookhaven National Laboratory
Idaho National Engineering Laboratory
Lawrence Berkeley Laboratory
Los Alamos National Laboratory
Oak Ridge National Laboratory
Sandia National Laboratory

LEWIS PUBLISHERS, INC.
121 South Main Street, Chelsea, Michigan 48118

ISBN 0-87371-417-2
Second Printing, 1991

Cover diagram: Global change is a highly intercoupled issue, linking human activities, emissions, concentrations, climate, and ecosystems and resources — each very broadly interpreted. Human activities include transportation, electricity generation, agriculture, heating and cooling, and more; emissions of concern include carbon dioxide, methane, chlorofluorocarbons, and more; concentrations are affected by land, oceanic and biospheric interactions, scavenging, chemistry, and more; climatic effects encompass changes in temperature, precipitation, sea level, and more; and ecosystems and resources affected include forests, water supplies, coastal and marine habitats, agriculture, and more.

TABLE OF CONTENTS

PREFACE

To fulfill his commitment to develop "a clear energy blueprint to take the U.S. into the next century," President George Bush commissioned Department of Energy Secretary Admiral James D. Watkins to develop a National Energy Strategy (NES). The Department of Energy national laboratories have been asked to assist in this effort by analyzing several key issues. This report has been prepared to provide an overview of scientific understanding of the potential chemical, climatic, and environmental effects of continuing emissions of carbon dioxide and other radiatively and chemically active trace gases and to balance the need for a concise statement with the need for consideration of the many complex aspects of this issue. The report seeks to emphasize the aspects on which there is broad scientific consensus, at the same time recognizing that important uncertainties remain. Where differing estimates are emerging from quantitative scientific studies, these are identified, but there is no attempt, nor is it intended, that this report be a complete and comprehensive review or that it address the sometimes wide array of opinions regarding the greenhouse effect on an individual basis.

Through the emission of greenhouse gases, the climate change issue is intimately tied to meeting the energy demands of the United States and all of the nations of the world. It is essential that we understand the degree of our scientific understanding, the rate of progress we can make in reducing uncertainties, and the commitment to further change that is occurring as research continues. This report, however, is not intended as a departmental, national, or international research plan, although it does identify issues whose resolution is needed to advance scientific understanding of the issue. Beyond considerations of scientific understanding are questions of policy recommendations; the development of energy scenarios, cost and risk analyses, and the effectiveness and appropriateness of various policy responses to explore these issues are not covered in this report, but are to be considered later in the NES process.

An earlier version of this report has been circulated for review by the Department of Energy and to scientists competent in the areas covered. A number of helpful suggestions have been provided, many of which have been incorporated. The issue, however, is very complex and time constraints have been rather limiting. Consequently, the authors view this report as a dynamic document, one that should and will be updated in the future in a continuing effort to provide the best possible technical evaluation for use in developing and maintaining the National Energy Strategy. Comments on this report would, therefore, be welcomed by the authors through any of the Department of Energy national laboratories.

Michael C. MacCracken
Chair, Multi-Laboratory Climate
Change Committee
February, 1990

DOE MULTI-LABORATORY CLIMATE CHANGE COMMITTEE

This report has been prepared cooperatively by scientists from the Department of Energy national laboratories and from other cooperating institutions. Most of the contributors have been participants in the DOE's ten-year carbon dioxide research program. The asterisk (*) indicates those who serve as their laboratory's lead representative on the Climate Change Committee.

Michael MacCracken, chair

Eugene Aronson
David Barns
Sumner Barr*
Cary Bloyd*
Dale Bruns
Robert Cushman
Roy Darwin
Donald DeAngelis
Michael Edenburn
Jae Edmonds*
William Emanuel
Dennis Engi*
Michael Farrell*
Jeremy Hales
Edward Hillsman
Carolyn Hunsaker
Anthony King
Albert Liebetrau
Bernard Manowitz*
Gregg Marland
Sean McDonald
Joyce Penner
Steve Rayner
Norman Rosenberg
Michael Scott*
Meyer Steinberg
Walter Westman*
Donald Wuebbles
Gary Yohe

Affiliations are given in Appendix 1.

We are indebted to Ms. Julia Bagorio and Ms. Mabel Moore for preparing the manuscript, to Ms. Mary Ann Esser and Ms. Anne Poore for editorial reviews, and to many others who reviewed and assisted in preparation of this report.

ABSTRACT

The atmospheric concentrations of carbon dioxide and other radiatively and chemically active trace gases have been increasing as a result of energy generation, transportation, and industrial, agricultural, and other societal activities. The carbon dioxide concentration has increased about 25% over preindustrial levels, and the methane concentration has more than doubled. Chlorofluorocarbon emissions have apparently initiated reductions in stratospheric ozone. Our understanding of the relationships of past emissions to past changes in concentration is developing rapidly, although it is still inadequate to satisfactorily explain important questions concerning the relative importance of oceanic and biospheric carbon reservoirs and the interactions of chemistry and climate on atmospheric concentrations. For scenarios for which arbitrary controls are not imposed, the major uncertainty in estimating future increases in the concentrations of these species results from uncertainties in estimating future emissions from energy, agriculture, and other activities. For scenarios where emissions are specified or limited, understanding of the fluxes into and out of the biosphere and oceans contribute the largest uncertainties. As a consequence, estimates of the time when carbon dioxide doubling will occur span much of the next century beyond the year 2030. Reduction of fossil-fuel emissions would slow the buildup of carbon dioxide in the atmosphere and allow additional time for adaptation to the changes, but modeling studies suggest that natural processes may not be sufficient to reduce the atmospheric concentration unless releases from human activities are reduced by 50 to 80% or more.

Models that simulate present and past climatic behavior suggest that an increase in carbon dioxide concentration to double the preindustrial level would commit us to an increase in global average temperatures of up to a few degrees Celsius. This warming would be accompanied by many other related changes of climate. An increase in annual average temperature by such an amount, while not unprecedented on a local basis, would elevate global temperatures to levels not experienced in historical times. This climate change could occur before the carbon dioxide concentration doubled because of the similar radiative effects of other gases, necessitating coupled consideration of the climatic and chemical effects of carbon dioxide and other trace gases, including especially methane, chlorofluorocarbons, and nitrous oxide.

On regional scales and for changes in precipitation, soil moisture, and important parameters other than temperature, the model simulations do not yet show strong agreement on the pattern or timing of the expected changes. Nor can models yet estimate the changes in frequency and intensity of precipitation systems and low-frequency events (e.g., tropical storms, droughts), but such changes seem likely to occur. The global warming evident over the past hundred years is approaching an amount beyond that which is thought to be due to contributions of natural variability over this period, suggesting that the greenhouse warming is under way. However, neither model calculations nor comprehensive observations are yet sufficient to establish a convincing and quantitative cause-effect relationship.

Although there are uncertainties in projecting future emissions, concentrations, and climatic effects, projections of possible changes are of sufficient magnitude and reliability to justify consideration of potential impacts on health, water supplies, agriculture, ecosystems, coastal areas, and other resources. Initial investigations into the sensitivities and vulnerabilities of these resources to composition and changes of climate, especially storms and other low-frequency events, suggest that many, but not all, resource impacts will be negative, particularly if the rate of climate change accelerates above its recent rate of about 0.5°C per century. Especially vulnerable are coastal areas subject to sea level rise, which may amount to as much as a half to one meter by the year 2100, continental interiors where warming is likely to increase the stress on increasingly limited water resources, and ecosystems, such as forests, that may not be able to migrate as rapidly as climatic zones shift. There may also be changes in the level and seasonal pattern of energy demand and supply that would require changes in the energy infrastructure.

The issue of climate change is global in extent. Although the U.S. must contribute to moderation or mitigation of the greenhouse issue, there is neither a U.S. solution nor a single technological adjustment or control measure that could readily and economically limit the increase in the concentrations of carbon dioxide and other trace species. A broad-based strategy will have to be developed that incorporates an array of options. Society can select from a menu of possibilities, including controlling emissions, changing technologies, and coping with the changes, even to the extent of attempting to counteract them. Choices among these options should be considered in the context of the broad range of social, economic, environmental, and other societal objectives. Capturing and sequestering carbon dioxide as it is produced in power plants and other combustion sources appears to have high technical and economic penalties. A combination of strategies that includes increased energy efficiency, more extensive use of alternative energy sources, reconstituted carbon fuels, and intensive reforestation could help slow the increase in carbon dioxide concentration in the intermediate term, but drastic reductions in CO_2 emissions would be required to come near to stopping the growth in the atmospheric CO_2 concentration. A range of other options exists to slow emissions of other trace gases.

Over the long term, an important technological challenge for the U.S. and the world is to develop and improve technologies for generating and utilizing energy that can improve the global standard of living and not create significant climatic and environmental impacts. It will likely take a broad-based energy research program and a range of incentives to induce necessary adoption and achievement of both the intermediate and long-term strategies.

There is no doubt that a wide range of activities generally highly beneficial to mankind is changing the atmospheric concentrations of radiatively and chemically active gases. Empirical evidence from past climates on the Earth and from planetary atmospheres reenforces the findings of numerical models that these changes in atmospheric composition can and will alter the climate and stress the living environment. Scientific studies cannot yet be certain in their projections of the precise amount, timing, or pattern of the climatic and environmental responses, but experience from climatic anomalies and past changes in climate suggest that the societal impacts of continuing emissions could be significant. Consideration of this issue in the development of the National Energy Strategy will require addressing a series of very difficult and interconnected questions concerning:

- Projection of future energy use and the mix of technologies.
- Estimation of the consequent level of chemical, climate, and environmental change.
- Selection of the acceptable rate of chemical, climate, and environmental change.

- Estimation of the probable rate of conservation and of technological and efficiency developments.
- Determination of the appropriate level of national and international commitment for addressing this global issue.

This report attempts to summarize the technical basis for addressing these issues, even while recognizing the many complex societal aspects underlying each of them.

DETAILED TABLE OF CONTENTS

Chapter 1: SUMMARY

1.1 The Changing Atmospheric Composition

The climate of the Earth provides the context for human activity, the suitable environment for agriculture and forests, and the sustenance for nature. Climate has varied dramatically over the past, from glacial conditions last peaking 18,000 years ago to much warmer conditions last present several million years ago. These changing conditions are controlled over the long term by changes in solar irradiance, in the shape of the Earth's orbit, by the height and extent of the continents, and, of present concern, by the composition of the Earth's atmosphere.

For the thousands of years before the industrial revolution, human activities affected only local atmospheric conditions. Smoke and carbon monoxide caused localized health impacts but did not affect the global environment. Wood provided the primary energy service; its combustion temporarily returned carbon dioxide (CO_2) to the atmosphere, but the carbon was soon captured again by the regrowth of the biosphere. Human activity was modest in extent. Measurements of atmospheric composition in air bubbles trapped in glacial ice indicate that concentrations of carbon dioxide remained stable over historical times.

The climate determined by this stable atmospheric composition has also remained relatively steady. Reconstructions of temperatures over the globe based on geological and ecological techniques suggest that, since the time when the last major glacial advance ended more than 10,000 years ago, global-average variations from the long-term mean have generally been less than about 1°C. Although regional variations in temperature have been somewhat larger, climate model simulations suggest that these changes have been due largely to the redistribution of solar energy caused by slow and cyclic changes in the Earth's orbit around the Sun. It is these orbital variations, in association with natural changes in atmospheric composition and/or ocean circulation, that are believed to be the primary driving forces for the glacial cycling over the past million years. Over the past few thousand years, the climate has been relatively stable, with decadal-average climate variations generally less than 1°C whereas in earlier epochs, glacial climates had been about 4°C colder and nonglacial epochs 5 to 10°C warmer. As a result of this quite stable climate, the distributions of peoples and the infrastructure for sustaining civilization have evolved in an environment characterized by a relatively stable climate.

The distributions of peoples and the infrastructure for sustaining civilization have evolved in an environment characterized by a relatively stable climate.

With the industrial revolution came the rapidly increasing use of energy and the rapidly spreading conversion of forests to agricultural fields. The energy was derived primarily from carbon that tens of millions of years ago and more had been removed from active exchange within the atmosphere-ocean-biosphere system by its burial as fossils. Emission of fossil-fuel-derived carbon dioxide as an energy waste product has been a major contributor to the 25% increase in the atmospheric carbon dioxide concentration since 1800, a result that has been confirmed by the observed increase in the CO_2 concentration in ice-core bubbles and in the dilution of the atmospheric concentration of the carbon-14 isotope.

The extension of agriculture to feed the increasing global population has contributed

By comparing the Earth's climate to that of neighboring planets, scientists have determined that the atmospheric composition acts in conjunction with planetary albedo and the distance from the Sun to control surface temperature.

to forest destruction and humus oxidation, moving carbon from storage in the biosphere to gaseous form in the atmosphere, thereby exacerbating the fossil-fuel effect. Present estimates are that annual fossil-fuel emissions exceed 5 billion metric tons of carbon, and net annual carbon release from deforestation exceeds 1 billion metric tons of carbon. The two effects together, even after uptake of some of the carbon dioxide by the oceans, are now increasing the atmospheric carbon dioxide concentration by more than 0.5% per year. Projections indicate that increased use of coal, petroleum, and natural gas will lead to an accelerated increase in the carbon dioxide concentration.

1.2 The Greenhouse Effect

Like oxygen and nitrogen, which make up most of the Earth's atmosphere, carbon dioxide lets most of the solar radiation pass through to warm the surface of the Earth. Unlike oxygen and nitrogen, however, the carbon dioxide molecule is quite effective at absorbing the infrared (heat) radiation emitted by the Earth's surface. Even though carbon dioxide currently makes up only 0.035% of the number of atmospheric molecules (350 parts per million by volume, ppmv), the absorption and re-emission of infrared radiation by carbon dioxide and water (both as vapor and in clouds) are very large. As a result, the atmosphere emits to the Earth's surface almost as much energy each day as strikes the atmosphere from the Sun and about twice as much as the solar radiation that makes it through the atmosphere to be absorbed at the surface (see Figure 1.1).

The atmosphere's near-transparency to visible radiation and strong trapping of infrared energy is often referred to as the *greenhouse effect*; although the atmosphere acts somewhat differently than do the panes of glass of a greenhouse, the effect is similar. The net consequence of the interception of infrared radiation by the *greenhouse gases* is to transform the climate of our planet from the harshness characteristic of lunar conditions to the more equable conditions of Earth. Estimates are that the greenhouse effect raises the surface temperature by about 33°C above the sub-freezing temperature the Earth would have without an atmosphere (assuming the same planetary reflectivity).

By comparing the Earth's climate to that of neighboring planets, scientists have determined that the atmospheric composition acts in conjunction with planetary albedo and the distance from the Sun to control surface temperature. Thus, although Venus is closer to the Sun than is the Earth, the bright clouds of Venus reflect so much sunlight that less energy is absorbed per square meter than by the Earth; nevertheless, because of the very high concentration of greenhouse gases in the atmosphere of Venus, its surface temperature is about 500°C compared to the Earth's average temperature of 15°C. Conversely, Mars is very cold, not only because it is farther from the Sun but also because its atmosphere lacks water and therefore the ability to trap infrared radiation. Geological studies have shown that during the Cretaceous, the epoch in Earth's history when dinosaurs roamed the continents and the period when much of the fossil-fuel resources were being formed, global average temperatures were as much as 10°C warmer than at present (Budyko et al., 1987). Although factors such as different continental locations also played a role, this was almost certainly in large part because the atmospheric carbon dioxide concentration was several times higher than at present.

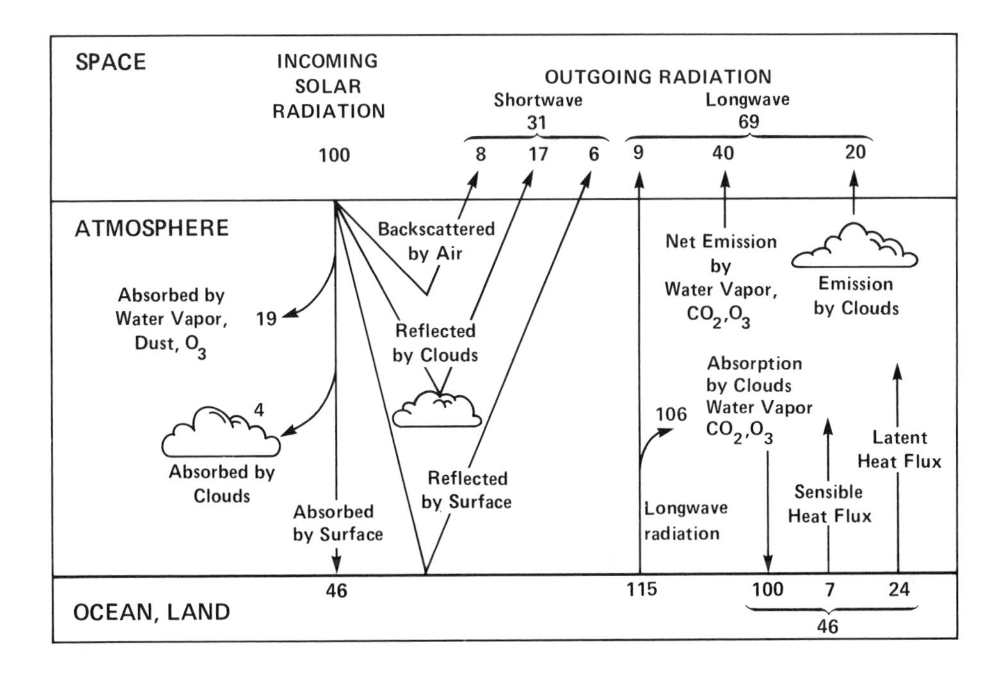

Figure 1.1 Schematic diagram of the relative sizes of the global average components of the Earth's energy balance expressed in terms of percent of the incoming solar radiation at the top of the atmosphere. Of the 100 units of incoming solar radiation (each equivalent to about 3.4 W/m^2), about 23 are absorbed in the atmosphere, 46 are absorbed by the surface, and 31 are reflected to space. The planetary energy balance is achieved by the emission of 69 units to space as infrared (longwave) radiation, mostly from the atmosphere and clouds. The solar energy absorbed by the surface is used in part to directly heat the atmosphere (sensible heat flux) and to evaporate moisture (latent heat flux). Atmospheric emission of infrared radiation downward to the surface is about equal to the solar radiation reaching the top of the atmosphere and more than twice as large as the amount of solar radiation absorbed at the surface. This greenhouse energy permits the surface to warm significantly more than would be permitted by the solar radiation alone (from MacCracken and Luther, 1985a).

Emission of carbon dioxide is not the only way in which human activities are altering the atmospheric concentrations of gases that can intensify the greenhouse effect. Since about 1800, the atmospheric methane concentration has more than doubled, scientists believe primarily as a result of emissions from increasing cultivation of wetland rice, increasing numbers of ruminant animals (cattle and sheep), and leakage from natural gas systems, transport, and use.

Chlorofluorocarbons (and halocarbons) were developed as inert, nontoxic substances and have found uses as refrigerants, foaming agents, degreasers, aerosol propellants, and fire retardants. Chlorofluorocarbon concentrations started at zero early in this cen-

The alteration of both atmospheric chemistry and radiative fluxes by the increasing emissions of various gases from combustion and other human activities significantly complicates the projection of future changes of climate.

tury, but these gases now are present in the atmosphere at a concentration of about one part per billion by volume (ppbv) and have raised the stratospheric chlorine concentration to above 2.7 ppbv (roughly four to five times natural levels). This buildup has occurred because these species are only slowly removed from the atmosphere by photochemical destruction high in the stratosphere. These destruction processes release chlorine and bromine atoms which, before their eventual removal in the lower atmosphere by precipitation, participate in catalytic reduction of the stratospheric ozone concentration. As a result, international action has been initiated to sharply curtail chlorofluorocarbon production, although sources are highly uncertain at present.

Nitrous oxide (N_2O) is another gas that can exacerbate the greenhouse effect and contribute to stratospheric ozone depletion. Its concentration is increasing as a result, it is suspected, of emissions from combustion and from soil microbes acting to break down the nitrogenous fertilizers that have proven so important in increasing global food production, although sources are highly uncertain at present.

Each of these gases has a direct effect on the atmospheric radiation balance by absorbing and re-emitting infrared radiation. At present concentrations, on a molecule for molecule basis, methane (CH_4) is about 30 times as effective as CO_2 and chlorofluorocarbons (e.g., CF_2Cl_2, $CFCl_3$, collectively referred to as CFCs) are up to 25,000 times as effective as CO_2 at trapping infrared radiation and warming the climate. Despite these ratios, there is so much more CO_2 than CH_4 and CFCs being added to the atmosphere, that the change in CO_2 concentration is now causing a perturbation to the radiation balance about equivalent to that of all of the other gases combined.

In addition to the direct effects that these greenhouse gases have on the fluxes of atmospheric infrared radiation, many of these gases—and others, including carbon monoxide (CO), hydrocarbons, and nitrogen oxides (NO and NO_2) generated by energy-related and industrial activities—can alter the natural atmospheric concentration of ozone (O_3), itself a greenhouse gas, and of hydroxyl (OH). This modification to the O_3 and OH concentrations can happen directly by altering the rates of chemical formation and destruction (e.g., in the case of nitrogen oxides) or by altering concentrations of catalytic reactants (e.g., by increasing the stratospheric concentration of chlorine). Changing the O_3 and OH concentrations can, in turn, cause changes the concentrations of CH_4 and other greenhouse gases. These greenhouse gas emissions can also act indirectly by changing the local temperature and thus the rate of reaction; for example, the stratospheric cooling induced by the increased CO_2-induced emission of infrared radiation slows the ozone-destroying catalytic reactions, leading, were other effects not acting, to an ozone increase. The alteration of both atmospheric chemistry and radiative fluxes by the increasing emissions of various gases from combustion and other human activities significantly complicates the projection of future changes of climate.

Changes in atmospheric composition change the fluxes of radiative energy in the atmosphere. To estimate the changes in radiative fluxes from the changing concentrations of greenhouse gases requires knowledge both of the radiative characteristics of these gases and of the atmospheric structure. Numerical, computer-based models incorporating the physics of atmospheric radiation have been developed for these calculations. These models have been tested by means of studies in the laboratory and intensive field observation programs. These tests indicate that the models have both strengths and weaknesses, the latter arising mainly from the difficulties of treating the radiative effects of clouds and the water vapor continuum part of the spectrum. When these models are used to estimate the effects of doubling the preindustrial atmospheric concentration of carbon dioxide projected to occur in the latter part of the 21st century, they indicate that there would be an increase in the initial trapping of infrared radiation in the surface-troposphere system by about 4 W/m^2 ($\pm$ 20%). The radiative effects of other greenhouse gases can be viewed either as an approximate doubling of this change or as advancing its occurrence by several decades.

1.3 The Climatic Response

A flux change of 4 W/m^2 is roughly equivalent to lighting a 15-W light bulb every 4 m^2 over the Earth.[1] To understand why this seemingly small change in atmospheric radiation should raise questions about altering national (and global) energy strategies requires understanding the interacting processes governing the global climate.

The climate system includes the atmosphere, oceans, land surface, cryosphere (glacial, mountain, and sea ice), and some aspects of the biosphere (Figure 1.2). These components interact to determine the climate. The *weather*, which is the instantaneous state of this system, is quite variable and turbulent; useful predictions of its chaotic and random behavior are possible only a few days in advance. The *climate*, which is the average of the weather over many years (by convention, 30 years), is much more stable (e.g., winter is always colder than summer). The dependence of climate on solar radiation, atmospheric composition, and other such separately predictable factors, offers hope that the effects on future climatic conditions of changes in composition can be estimated well into the future, even if weather can be forecast only a few days into the future.

The greenhouse question is not whether atmospheric composition is a primary determinant of the global average temperature and climate but, rather, how large and how fast the changes will be as atmospheric composition is altered.

The climate system is more complex than can be fully represented by a laboratory experiment, and there are no suitable climatic analogs from the past that match the projected extents and rates of change of composition. Because these more traditional approaches are not adequate, numerically-based computer models are the most suitable means for studying climate and climate change; the most comprehensive of these are referred to as general circulation models—GCMs. These models, which can represent many, though by no means all, of the most important of the interacting processes constituting the climate system, have undergone a wide, but still a limited, range of verification tests against observations. These tests suggest that the models can represent planetary-scale climatic features with reasonable fidelity, but regional-scale climatic features are represented with much less certainty, in large part because their spatial resolution is limited to several hundred kilometers. Tests of the GCMs also suggest that they can represent reasonably well short-term changes of climate, such as the seasonal cycle, and long-term equilibrium

[1] For reference, the annual average intensity of solar radiation reaching the surface is only about 170 W/m^2.

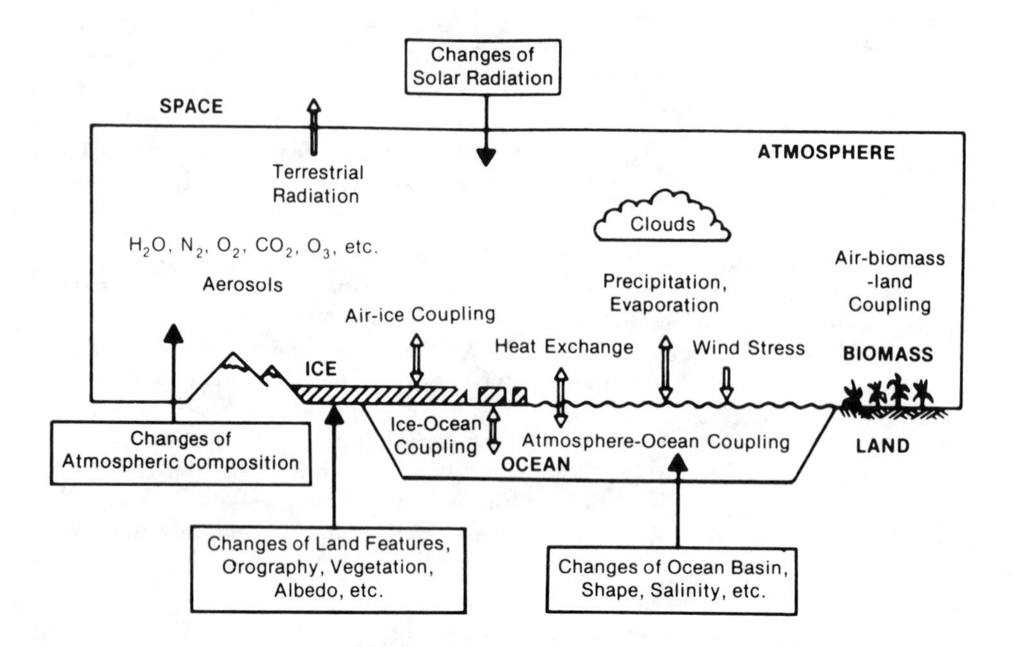

Figure 1.2 Schematic illustration of the components of the coupled atmosphere-ocean-ice-land climatic system. The solid arrows represent processes that force the climate system and the open arrows represent processes that govern and change in response to the forcing (Gates, 1979).

The climate models suggest that a doubling of the carbon dioxide concentration, or a radiatively equivalent increase in the concentration of all greenhouse gases, would lead to an increase in global average temperature of 1.5 to 4.5°C, once the oceans warmed and a new equilibrium climate was established.

changes of climate, such as glacial-interglacial conditions. However, climate models are largely untested against the decadal scales of interest in global climate-change studies, principally because the causes of changes of climate on these scales remain poorly understood.

That the climate is warmed by greenhouse gases is certain. The greenhouse question is not whether atmospheric composition is a primary determinant of the global average temperature and climate but, rather, how large and how fast the changes will be as atmospheric composition is altered. The need to understand the potential effects on climate—and thus on society—has led to use of the GCMs, incomplete as they may be, to estimate the sensitivity of the climate to changing composition.

The climate models suggest that a doubling of the carbon dioxide concentration, or a radiatively equivalent increase in the concentration of all greenhouse gases, would lead to an increase in global average temperature of 1.5 to 4.5°C, once the oceans warmed and a new equilibrium climate was established.

Associated with this increase would be other changes, perhaps including a significant extension of the warm, ice-free season at high latitudes, intensified drying of continental interiors in summer (except perhaps in monsoon dominated regions), elevated snow levels (perhaps to more winter and less summer runoff in mountainous areas), and increased likelihood of more intense tropical storms resulting from a greater areal extent of ocean waters warmer than about 27 or 28°C. The projection of these aspects, especially changes of climate in particular regions, is, however, much less certain than the forecast of an overall global warming.

The rate of emission of the greenhouse gases (which determines the evolving composition of the atmosphere), the sensitivity of the Earth's climate to changes in greenhouse trapping of infrared radiation, and the rate of heat uptake by the ocean will together determine the rate of climate change. The rate of ocean warming depends, in turn, on the coupling of the relatively low heat capacity of the upper ocean to the large heat capacity of the deep ocean. Model simulations that take the deep oceans fully into account are only beginning to be available. Early results not only indicate that the time-dependent climate change will lag the equilibrium change by a few to several decades but also suggest that the spatial patterns of the time-dependent climate change are significantly different from those obtained by interpolating between the equilibrium climatic calculations that have been carried out so far. Over the next several decades, an average global warming rate of up to about 0.2 to 0.4°C per decade is suggested by the ensemble of model results. The range of estimates results in large part from uncertainties in how cloud cover will change in response to the warming.

Examination of the instrumental record of the past 150 years offers the potential to

Over the next several decades, an average global warming rate of up to about 0.2 to 0.4°C per decade is suggested by the ensemble of model results. The range of estimates results in large part from uncertainties in how cloud cover will change in response to the warming.

search for evidence that the warming has begun as a result of the increases in concentration of CO_2, CH_4, and other gases that have already occurred. Unfortunately, problems with measurement techniques, limited areal coverage of the set of observing stations, and other factors have limited and contaminated the data sets such that accuracy over the hundred-year period is probably limited to a few tenths of a degree. In addition, although the global climate of the past several thousand years has been stable to within perhaps one degree, natural factors predating significant industrialization have apparently caused climatic variations of several tenths of a degree lasting up to a few centuries, particularly around the North Atlantic basin where our observational (and proxy) climate records are best. These complications, among others, make quantitative comparison of the approximate 0.4 to 0.5°C rise in global average temperature over the past 100 years with the model simulations of greenhouse warming somewhat problematic.

Although the observed rise in temperature over the past 100 years seems unusual, indeed almost unprecedented given our understanding of natural variations, we cannot yet confirm that model projections are valid, at least not to within about a factor of 2. Until we have better estimates from the models and confirmation from observations of the patterns of climate change, we can only infer that the greenhouse effect is increasing. We are still short of being able to explain precisely and determine quantitatively the cause-effect relationship.

Although the observed rise in temperature over the past 100 years seems unusual, indeed almost unprecedented given our understanding of natural variations, we cannot yet confirm that model projections are valid, at least not to within about a factor of 2.

1.4 Environmental Response to Climate Change

Climatic conditions are also tied to the natural terrestrial and marine biosphere in complex ways of varying strength and duration. On the decade-to-centuries time scale of the greenhouse gas effects, the biospheric response may moderate the changes slightly (e.g., by enhanced biospheric growth and by increased uptake of carbon dioxide as the concentrations rise). However, many of the changes that have been identified tend to amplify the perturbation (e.g., see Lashof, 1989) rather than to slow or ameliorate the projected changes of climate. Rather, the changing climate seems likely to alter significantly the environment in which present biospheric, agricultural, hydrologic, and economic systems have developed.

We are only starting to glimpse the range and complexity of possible environmental consequences as the various dependencies and interdependencies of climate and ecosystems become better understood. Given that the activities generating greenhouse gases, especially energy consumption, provide substantial benefits to mankind, the search for potential impacts should focus on areas where countervailing consequences may be of comparable dimension. Thus, initial examination is focusing on impacts on human health, food and fiber resources, water resources, coastal structures and shoreline resources, and the diverse natural biological environment.

Although individuals who move from Maine to southern California survive much greater climate and environmental changes than are projected from the rising concentrations of greenhouse gases, many plant and animal species cannot so readily adapt. For those individuals who do not move, dealing with the health effects of an increased frequency of warm and hot days would require adjustments, especially in hot and humid climates. Indirect effects of the changes of climate affecting health may well include increased prevalence of pests and other disease vectors and exacerbation of summertime energy demands. Although warmer winters may allow emissions to be reduced without a health impact, curtailment of the emissions resulting from energy consumption and air conditioning now providing human comforts may well impose health impacts, that, depending on the investment level for new technologies, may be as large as the health-related impacts induced by the emissions and consequent changes of climate.

Because carbon dioxide is necessary for plant growth, the increase in its concentration might cause some nonwoody plants to grow more rapidly and with improved water-use efficiency, presuming other factors controlling plant growth (e.g., soil nutrients) are not limiting. Studies indicate, however, that the nutritional value of plants might diminish, that growth enhancement depends on many other factors, and that many weeds seem to respond more positively to an increase in carbon dioxide than do many crops. Furthermore, the long-term effects of an increased CO_2 concentration on woody plants (trees and shrubs) are unknown. Were there no infrastructure and competitive costs for crop switching and relocation, technologically advanced and evolving agricultural systems might benefit from a warmer world, but there are serious constraints, especially limitations on water availability and soil characteristics,

We are only starting to glimpse the range and complexity of possible environmental consequences as the various dependencies and interdependencies of climate and ecosystems become better understood.

that might lead to serious adverse consequences.

Changing precipitation patterns, a higher snowline, and increased evaporation seem likely to increase significantly stresses on available water resources. Although model projections are much less certain about the hydrologic response, especially on regional scales, increased summertime temperatures probably will exacerbate the dry climatic fluctuations now occurring in some regions, making drought conditions more frequent and intense. At the same time, increased frequency of tropical storms might lead to more frequent, although still irregular, inundation of susceptible regions.

A rise in sea level, which initial projections indicate may amount to as much as half to one meter by the year 2100, and more thereafter, would threaten many coastal areas, especially during storm surge conditions. While dikes and levees may be cost effective for protecting highly developed areas, inundation of wetlands and low-lying coastal plains (e.g., Louisiana, Florida, Bangladesh) and coral islands could be extensive.

Studies of ecosystem composition and distribution under the differing climatic conditions of the past indicate that life is possible under these conditions but also suggest that ecosystems could be very different and individual species could be highly disrupted. For example, species of plants and animals may not be able to migrate to keep up with the shifting climatic conditions, with consequent destruction of forests and reduction of biodiversity. Recent simulations of

Studies of ecosystem composition and distribution under the differing climatic conditions of the past indicate that life is possible under these [perturbed] conditions but also suggest that ecosystems could be very different and individual species could be highly disrupted.

forest growth indicate the potential response of ecosystem productivity and biomass to climate warming may depend on soil water availability and feedback regarding nutrient cycling. Also, climate-induced disturbances like drought (and resultant wildfire) may be favored by greenhouse warming, accelerating vegetation change and altering biomass at rates faster than that due to global warming alone.

This set of effects seems to be only the tip of the iceberg. As the social and economic systems tied to the health, food, water, coastal, and natural systems change, further consequences will be induced, such as movement of societal activities into less robust ecological areas (e.g., into current permafrost regions). The complexities of inquiring into these effects are formidable, but research on the entire chain of steps is possible—and vital—if we are to be able to carry through more comprehensive analyses.

1.5 *Technological Options for Responding*

Just as it is a significant challenge to understand the chain of effects from emissions to concentration changes to changes of climate and environmental responses, developing a range of technological options for addressing the issue poses a significant challenge. General classes of possible options include reducing emissions, producing and using fossil fuels more efficiently, switching to

In view of the increasing global population, the universal demand for a higher standard of living, and the limitations posed by finite resources, global changes of climate challenge us to rethink our energy strategy and technologies, our use of resources, and our interactions with and responsibilities to the natural environment.

nonfossil-fuel technologies, attempting to intervene and modify the global climate, adapting (perhaps actively) to the changes of climate and consequences, or some combination of these. Specific actions might include more efficient supply and use of energy, reforestation, switching from coal to natural gas, increased emphasis on nuclear power, or fertilizing the oceans. While some of the possible actions may be marginally advantageous at present, substantial slowing of the rate of greenhouse gas emissions would require a substantative commitment of resources (and perhaps a significant slowing of the international effort to raise the global standard of living) unless nonfossil-fuel-based energy sources become available at prices generally consistent with those for fossil-fuel resources.

In view of the increasing global population, the universal demand for a higher standard of living, and the limitations posed by finite resources, global changes of climate challenge us to rethink our energy strategy and technologies, our use of resources, and our interactions with and responsibilities to the natural environment. Virtually all of the greenhouse gas emissions of concern result from activities that provide substantial benefits, including energy, food, and health maintenance. There is no single or simple technological fix that can suddenly transform our "industrial ecosystem" and eliminate these emissions. Stabilizing emissions at current levels will not stop future increases in the atmospheric concentration of greenhouse gases, even though it will prevent the further acceleration of such changes; even stabilizing atmospheric composition, which would require drastic reductions in current emission levels, would not stop some further climatic warming as a result of past emissions. But without any controls, the rate of climate change will accelerate so that by the middle of the next century the global temperature will reach a level significantly greater than exists today or than global temperatures experienced in historical times. To slow the future rate of global climate change will require a global strategy that is supported and pursued by all nations, each doing its part.

1.6 Overview of Report

Chapters 2 through 6 of this document deal with greenhouse gas emissions, atmospheric composition, climatic response, human implications of potential climate changes, and possible societal responses to climate change. The objective of each of these chapters is to describe the current scientific understanding and attendant knowns, unknowns, and uncertainties associated with the global climate issue. Thus, each chapter summarizes and points out limitations in present understanding. Chapter 7 then summarizes the findings that would contribute to more comprehensive information for the public and for informed policy development through the National Energy Strategy.

The material that is covered is drawn from the full range of international scientific research on this subject, but it draws especially from the series of state-of-the-art reports prepared under the auspices of the Department of Energy's Carbon Dioxide Research Program (Trabalka, 1985; MacCracken and Luther, 1985a, 1985b; Strain and Cure, 1985) and the draft of the potential environmental effects prepared by the EPA (Smith and Tirpak, 1988). Other reviews on behalf of the National Research Council (NRC,

1983) and the Scientific Committee on Problems in the Environment (Bolin et al., 1986) provide a similar picture. All of the reports indicate that there are significant limitations to current understanding but also that the prospects for future change are substantial and deserving of special consideration.

Chapter 2: EMISSIONS

2.1 Energy and Greenhouse Gas Emissions

The concern about climate change has arisen because human activities have expanded to the point that worldwide releases to the atmosphere of CO_2 and other radiatively important gases are changing the global atmospheric concentrations of these gases. This chapter examines emissions of gases that either directly or indirectly (through atmospheric chemistry) contribute to the greenhouse effect. Table 2.1, expanded from Wuebbles and Edmonds (1988) and Wuebbles et al. (1989), summarizes the characteristics of these gases and others that are released by human activities or whose atmospheric concentrations are climatically and/or chemically important and whose concentrations are affected by human activities. This includes especially ozone, which is a naturally produced greenhouse gas resulting from the emission of CH_4, CO, nitrogen oxide (NO_x), and other chemical interactions (see Section 3.3.4) including those with N_2O. Interactions between CO, volitale organic compounds (VOCs), OH, and CH_4 are also important.

Emission of these gases as a result of human activities takes place against a natural background of emissions and removal processes that range from nonexistent to enormous. Carbon dioxide is generally considered to be the most important of these gases, accounting for about half of the current and anticipated perturbation to the radiative forcing. Fossil-fuel use is the most important present and anticipated anthropogenic source of CO_2 emissions, although land-use change, especially deforestation, is another important net source to the atmosphere. In this chapter the emphasis will be on CO_2 and the other gases that are most important in affecting the global climate.

2.2 Historical Emissions of Greenhouse Gases

The quality and detail of estimates of past emissions of greenhouse gases vary considerably. Fossil-fuel CO_2 emissions and release of CFCs to the atmosphere are among the best documented of the greenhouse gas emissions; CO_2 emissions from land-use changes are less well understood. Table 2.2 characterizes the current understanding of historical CO_2 emissions. Emission budgets for CH_4 and N_2O are also uncertain. The global emission budgets of CO and NO_x, which affect climate indirectly by perturbing the atmospheric composition, are highly speculative. Fluxes of carbon between oceans and the atmosphere and between the biosphere and the atmosphere are much larger than emissions from fossil-fuel use and land-use change; these exchanges of carbon are treated in Chapter 3 as part of the discussion of the natural carbon cycle.

2.2.1 Historical Fossil-Fuel CO_2 Emissions

There is a longer record of emissions of CO_2 from fossil fuels than for any of the other gases. Since 1860, global annual emissions of fossil-fuel CO_2 have increased from 0.1 to approximately 5.7 PgC/yr in 1987.[1] Cumulative fossil-fuel carbon emissions over this period total almost 200 PgC. During the period 1945 through 1979 the rate of CO_2 emissions from fossil-fuel use grew at a rate of 4.5%/yr. Emissions declined from 1979 until 1983 and have risen subsequently. The U.S., U.S.S.R., and People's Republic of China (P.R.C.) account for about half of the world's fossil-fuel CO_2 emissions. U.S. fossil-fuel CO_2 emissions

[1] gC = grams of carbon; 1 PgC = 1 petagram of carbon = 10^{15}gC = 1 gigatonne C = 1 billion tonnes (metric tons) of carbon.

Table 2.1 Important trace gases affecting the composition of the atmosphere and climate.

Gas	Common name	Primary anthropogenic sources	Greenhouse gas?	Interactions[a] with tropospheric composition?	Interactions[a] with stratospheric composition?
CO_2	Carbon dioxide	Fossil-fuel burning; land-use conversion.	Yes	No	No
CH_4	Methane	Ruminant animals; rice paddies; biomass burning; gas and mining leaks.	Yes	Yes(OH,O_3)	Yes(O_3,H_2O)
CO	Carbon monoxide	Energy use; agriculture; biomass burning.	Yes, but weak	Yes(OH,O_3)	Not significantly
N_2O	Nitrous oxide	Cultivation and fertilization of soils; combustion (uncertain); deforestation (uncertain).	Yes	No	Yes(O_3)
NO_x ($=NO + NO_2$)	Nitrogen oxide	Fossil-fuel burning; biomass burning.	Yes	Yes(OH,O_3, NMHC)	Yes(O_3)
$NO_y(=NO_x + HNO_3 + PAN + N_2O_5)$	Reactive odd nitrogen	Primarily created by atmospheric reactions involving NO_x.	Yes	Yes(OH,O_3, NMHC)	Yes(O_3)
$CFCl_3$	CFC-11	Chemical industry.	Yes	No	Yes(O_3)
CF_2Cl_2	CFC-12	Chemical industry.	Yes	No	Yes(O_3)
$C_2Cl_3F_3$	CFC-113	Chemical industry.	Yes	No	Yes(O_3)
CH_3CCl_3	Methyl chloroform	Chemical industry.	Yes	Yes(OH)	Yes(O_3)

[a] Interactions as used here refers to direct chemical interactions. By influencing atmospheric radiative fluxes and temperatures, all greenhouse gases also indirectly affect the concentrations of other species by changing the rate of reaction, photodissociation rates, or other factors.

Table 2.1 Important trace gases affecting the composition of the atmosphere and climate (continued).

Gas	Common name	Primary anthropogenic sources	Greenhouse gas?	Interactions[a] with tropospheric composition?	Interactions[a] with stratospheric composition?
CF_2ClBr	Ha-1211	Fire extinguishers.	Yes	Yes (photolysis)	Yes(O_3)
CF_3Br	Ha-1301	Fire extinguishers.	Yes	No	Yes(O_3)
SO_2	Sulfur dioxide	Coal and petroleum burning.	Yes, but weak	Yes(OH, aerosols)	Yes (aerosols)
COS	Carbonyl sulfide	Biomass burning; fossil-fuel burning	Yes, but weak	Yes(OH)	Yes (aerosols)
DMS	Dimethyl sulfide	Primarily natural.	No	Yes (aerosols)	No
NMHC	Non-methane hydrocarbons	Incomplete oxidation of carbon.	No	Yes(O_3,OH)	No
O_3	Ozone	Not directly emitted, created by reactions of NO_x and NMHC in troposphere and by photolysis in the stratosphere.	Yes	Yes(many species)	Yes(many species)
OH	Hydroxyl	Not directly emitted, created naturally.	No	Yes(many species)	Not significant
H_2O	Water vapor	Anthropogenic emissions are small compared to natural evaporation.	Yes	Yes	Yes

[a] Interactions as used here refers to direct chemical interactions. By influencing atmospheric radiative fluxes and temperatures, all greenhouse gases also indirectly affect the concentrations of other species by changing the rate of reaction, photodissociation rates, or other factors.

Source: Based on Wuebbles and Edmonds (1988).

accounted for more than 40% of global emissions in 1950. This share has steadily declined to less than 25% in 1989. U.S. CO_2 emissions peaked in 1973 (1.27 PgC/yr) and again in 1979 (1.30 PgC/yr) and remained below that level through 1986 (1.20 PgC/yr), the last year for which global statistics are available. U.S. fossil-fuel CO_2 emissions have increased since 1983, however. Global average per capita emissions of fossil-fuel CO_2 to the atmosphere are approximately 1 tonne of carbon per year;[2] U.S. emissions exceed 5 tonne per person per year.

Carbon emissions to the atmosphere vary by fuel. For fossil fuels, natural gas is lowest (13.7 TgC/EJ),[3] coal is highest (23.8 TgC/EJ), and oil falls between the two (19.2 TgC/EJ). Dissociating the carbonate rock formations that contain oil shale would add an additional stream of CO_2 to the atmosphere, but this increment need not be large with certain techniques. The transformation of primary fossil-fuel energy, for example from coal to electricity or from coal to synoil or syngas, releases carbon in the conversion process. Energy technologies such as hydroelectric power, nuclear power, solar energy, and conservation (including energy efficiency improvements) emit no CO_2 directly to the atmosphere. It should be noted that more efficient fossil-energy technologies can also reduce greenhouse gas emissions relative to conventional technologies. Traditional biomass fuels, such as crop residues and dung, release CO_2 to the atmosphere but are generally in a closely balanced cycle of absorption and respiration, unless soil carbon is being depleted. The use of other biomass fuels such as firewood may provide either a net annual source or a sink for carbon, depending upon whether the underlying biomass stock is growing (sink) or being exhausted (source); left alone, forests are generally sinks of carbon.

[2] 1 tonne = 10^6 g = 1 Mg.

[3] TgC = 10^{12} gC; EJ = exajoule = 10^{18} J = 0.948 Quad.

Global average per capita emissions of fossil-fuel CO_2 to the atmosphere are approximately 1 tonne of carbon per year. U.S. emissions exceed 5 tonne per person per year.

The inclusion of indirect emissions of CO_2 to the atmosphere would alter these basic emissions coefficients. Indirect emissions of CO_2 would be encountered in all energy technologies. In the supply of nuclear power, for example, emissions occur in the mining of uranium and from the construction of the powerplant. Preliminary calculations indicate that the inclusion of indirect emissions does not change the relative emissions characteristics of energy carriers.

2.2.2 Historical CO_2 Emissions From Land-Use Change

There are approximately 600 PgC in terrestrial vegetation, principally in forests. This is estimated to be about 15–20% (~ 120 PgC) less than was present in the mid-nineteenth century. On a global basis, this is estimated to vary by less than about 10 PgC through the seasons as leaves and grasses grow and die.

Estimates of the net annual emissions of carbon from land-use changes are far less certain than estimates of fossil-fuel emissions. Emissions of net annual CO_2 release from land-use changes have been estimated for the year 1980 by various researchers. Net release is calculated as the difference between annual gross harvests of biomass, plus releases of carbon from soils, less biomass carbon whose oxidation is long delayed (e.g., stored in forest products such as telephone poles, furniture, and housing) and additions to the stock of

Table 2.2 Historical CO_2 emissions.

Sources	Quality of emissions data	Central estimate (Pg C/yr)	Year of estimate	Uncertainty	Notes
Fossil-fuel use	Good	5.7	1987	± 20%	Available for the years 1860 to 1987 by country and by fuel.
Cement manufacture	Good	relatively small	—	± 20%	
Land-use change	Fair/poor	0.4 to 2.6	1980	± 100%	Primarily tropical deforestation.

Source: Based on Wuebbles and Edmonds (1988).

standing biomass. Houghton et al. (1983) estimated 1980 land use emissions to be between 0.5 and 4.5 PgC/yr. This range has been narrowed by Houghton et al. (1987) to between 1.0 and 2.6 PgC/yr, but estimates remain quite uncertain. Detwiler and Hall (1988) estimate 1980 emissions from the tropics to be in the range 0.4 to 1.6 PgC/yr. Net emissions from land-use change are dominated by tropical deforestation. Houghton et al. (1987) estimate that all but 0.1 PgC/yr of net releases are from tropical forests. Estimates of deforestation in 1980 are greatest for Brazil, Columbia, the Ivory Coast, Indonesia, Laos, and Thailand.

Estimates of net CO_2 emissions from land-use change have increased for recent decades. Prior to 1950, significant deforestation is estimated to have occurred in the temperate latitudes as well as in the tropics. Although a matter of heated debate, it has been suggested that increases in the atmospheric concentration of CO_2 could act to accelerate the rate at which plants assimilate carbon. Estimates of net CO_2 release from land-use change have generally not taken the possibility of a CO_2 fertilization effect into account.

2.2.3 Historical Methane and Carbon Monoxide Emissions

Annual estimates of emission sources are not generally available for CH_4 or CO. Source strength uncertainties are so high that emission budgets are typically referenced to a decade average annual level rather than to an individual year. Whereas emissions of CO_2 from fossil-fuel and land-use change are developed for specific years on the basis of data bases for a small number of human activities, the sources and sinks for CH_4 must now be developed using an observed globally averaged atmospheric burden of CH_4 (approximately 4800 $TgCH_4$), an average annual rate of increase (approximately 1%/yr), and an atmospheric lifetime of approximately ten years derived from an atmospheric chemistry model. Together, these factors lead to an estimate of total global emissions of about 400 to 640 $TgCH_4$/yr. Information about the changing isotopic ratio of methane (i.e., $^{14}CH_4/^{12}CH_4$) is used to partition the emissions by broad period of origin and (by inference) to bound the contribution of fossil fuels to total emissions. A summary of the best current understanding of the sources and sinks of CH_4 is given in Table 2.3.

At this point it is not clear that all of the major sources of CH_4 have been identified, and the emission rates of those that have been identified are subject to significant uncertainty. Anthropogenic activities are currently thought to contribute approximately half of all CH_4 emissions to the atmosphere. The three principal human activities that have been identified as emission sources of methane are cattle raising, rice production, and energy production and use. Although human activities have been identified as major sources of atmospheric emissions, there remains great uncertainty surrounding emissions source estimates and the time profile of those emissions.

Roughly a quarter of the total atmospheric methane emissions are attributable to the production, transfer, conversion, and consumption of energy. These include the mining of coal and the gathering, transmission, distribution, venting, and flaring of natural gas. Landfill material representing the residue of the consumption process is a rich source of methane which is only very slightly exploited as a source of energy at the present time. Burning of biomass can occur naturally, as in forest fires, or can be initiated by human activity, as in the clearing of land for agriculture. Some fraction of the human contribution is from direct energy consumption such as the burning of fuel wood. Finally, each of the combustion processes associated with the conversion of fossil fuel to thermal energy may be attended by the emission of some quantity of methane, depending upon the constituents of the fuel, the temperature of combustion, and the efficiency of the process. Emissions from natural gas production, coal mining, and landfills currently appear to be more important sources of methane than combustion process byproducts. Recently, pressurized water (nuclear) reactors have been identified as a very small but increasing source of methane (Wahlen et al., 1989). The U.S., U.S.S.R., and P.R.C. are the largest sources of fossil-fuel methane emissions.

The three principal human activities that have been identified as emissions sources of methane are cattle raising, rice production, and energy production and use ... Roughly a quarter of the total atmospheric methane emissions are attributable to the production, transfer, conversion, and consumption of energy.

The emission of CO is of interest because it enters into atmospheric chemical reactions that can affect the concentrations of radiatively active gases. These interactions are especially important for CH_4 as CO reacts rapidly with OH and, by depleting available OH, can extend the average lifetime of CH_4. Even greater uncertainty surrounds the atmospheric carbon monoxide (CO) budget than the CH_4 budget. As indicated in Table 2.4, CO is generated by incomplete combustion (complete combustion yields CO_2 rather than CO), oxidation of anthropogenic hydrocarbons, the decomposition of CH_4, and other minor sources. Because of its relatively short lifetime in the atmosphere (0.4 yr), the gas is poorly mixed in the global atmosphere. Estimates of annual emissions range from 600 to 1700 TgC/yr. Concentrations of CO are significantly higher in the Northern Hemisphere than in the Southern Hemisphere, indicating higher emissions in those latitudes. This is consistent with the pattern of combustion activities. U.S. emissions of CO from fossil-fuel use are estimated to have been approximately 24 TgC/yr in 1986. The most important sink for both CH_4 and CO is atmospheric chemical reaction leading to CO_2. The fossil-fuel component is generally already included in the

Table 2.3 Historical CH_4 emissions.

Sources and sinks	Quality of emissions data	Uncertainty	Notes
Sources			
Natural sources	Poor	Large	Total global emissions from natural and anthropogenic sources are estimated to range from 400 to 640 $TgCH_4$/yr. Natural sources are approximately 50% of global emissions. It is not clear that all sources and sinks for CH_4 have been identified. Total emissions are derived from atmospheric observations and calculations from atmospheric chemistry models. Current source estimates are (in $TgCH_4$/yr): enteric fermentation (wild animals), 4 ± 3; wetland, 110 ± 50; lakes, 4 ± 2; tundra, 3 ± 2; oceans, 10 ± 10; termites and other insects, 25 ± 20; methane hydrates, 5 ± ?; other, 40 ± 40.
Agriculture	Fair/poor	± 50%	25 to 40% of global emissions. Current source estimates are (in $TgCH_4$/yr): rice cultivation, 70 ± 30; ruminant digestive systems of domesticated animals (cattle), 77 ± 35; slash-and-burn agriculture/land clearing, 55 ± 30.
Energy	Fair/poor	± 50%	20 to 25% of global emissions. Current source estimates are (in $TgCH_4$/yr): Deep coal mining, 25 ± 20; natural gas production, transport, and distribution, 40 ± 15; incomplete combustion (e.g., automotive exhaust, fuel wood use), 15 ± 8; landfills, 30 ± 30.
Sinks			
Atmospheric chemistry	N/A	± 50%	Current sink estimates are (in $TgCH_4$/yr): reaction with tropospheric OH leading to CO_2, 350 ± 100; transport to and reaction with OH, Cl, or O in stratosphere leading to CO_2, 50 ± 20; uptake by soil micro-organisms, 32 ± 16; accumulation in the atmosphere, 60 ± 15.

Source: Based on Wuebbles and Edmonds (1988).

Table 2.4 Historical CO emissions.

Sources and sinks	Quality of emissions data	Uncertainty	Notes
Sources			
Natural sources	Poor	Great	Total global emissions from natural and anthropogenic sources are estimated to range from 600 to 1700 TgC/yr. Natural sources are approximately 50% of global emissions. Current source estimates are (in TgC/yr): plants, 55 ± 35; wildfires, 10 (5 to 20); oceans, 20 (10 to 40); oxidation of natural hydrocarbons, 250 (50 to 500); and oxidation of CH_4, 260 (75 to 450). It is not clear that all sources and sinks for CO have been identified. CO is highly reactive and therefore global source/sink relationships are highly uncertain even in aggregate.
Non-energy	Poor	± 100%	40 to 60% of global emissions. Current source estimates are (in TgC/yr): slash-and-burn agriculture/land clearing, 270 ± 250.
Energy	Fair/poor	± 50%	40 to 60% of global emissions. Current source estimates are (in TgC/yr): incomplete combustion (e.g., automotive exhaust, fuel wood use), 200 ± 100; oxidation of anthropogenic hyrocarbons, 40 ± 40. U.S. energy-related CO emissions have been estimated as part of the National Acid Precipitation Assessment Program (NAPAP).
Sinks			
Atmospheric chemistry	N/A	± 50%	Current sink estimates are (in TgC/yr): chemical reaction leading to CO_2, 820 ± 300.
Soil uptake	N/A	± 50%	Current sink estimates are (in TgC/yr): 110 ± 30.

Source: Based on Wuebbles and Edmonds (1988).

CO_2 emission budget; the biospheric components are generally less than 10% of the fossil fuel carbon and are accounted for in the biospheric CO_2 emissions inventory.

2.2.4 Historical Nitrous Oxide Emissions

The atmospheric concentration of N_2O is increasing at the rate of about 0.3%/yr (Watson et al., 1988; Khalil and Rasmussen, 1988). Current tropospheric concentrations are about 307 ppbv. Ice core data indicate that the concentration of N_2O was stable for approximately 3000 years at 285 ppbv. Concentrations began to increase beginning about 150 years ago.

The sources of N_2O emissions are poorly documented, as indicated in Table 2.5. Emission rates are small relative to atmospheric stocks. Although the atmospheric burden and annual rate of increase are known with some confidence, the atmospheric lifetime is uncertain, probably being within the range of 100 to 175 years. This uncertainty leads to significant uncertainties in estimating the total sources and sinks, which can be derived from atmospheric chemistry models. Individual source terms are subject to even greater uncertainty. Total emissions are estimated to be between 11 and 19 TgN/yr. Emission studies are inconsistent with regard to their categorization of emission-producing activities. The chief sources of emissions are presently thought to be biogeochemical decomposition in soils, and combustion activities. The biogeochemical processes include N_2O releases from cultivated and uncultivated soils and from fertilized and unfertilized soils. Combustion activities include savanna burning, forest clearing, fuel wood use, and fossil-fuel combustion. Other sources of emissions include oceans and contaminated aquifers.

The dominant human activities associated with N_2O emissions are agricultural (burning of grasslands, soil cultivation, and fertilizer application) and energy (wood burning and fossil-fuel use). Khalil and Rasmussen (1988) use ice core data to constrain the total human-related emissions to 10 to 30% of the total. Until recently, the dominant man-made emission source was thought to be fossil-fuel combustion (Hao et al., 1987; Wuebbles and Edmonds, 1988). This conclusion was based on flask samples taken from combustion experiments. This research was recently shown to be subject to a sampling artifact that produced N_2O in the flask between the time the sample was taken and the time the gas in the flask was analyzed (Muzio and Kramlich, 1988; Linak et al., 1989). It is possible that fossil-fuel emissions are a relatively minor source of N_2O emissions, but this is by no means certain. It is also possible that the chemistry that occurred in the flask may also occur in nature. Wuebbles and Edmonds (1988) estimate that approximately 24% of the N_2O sources are human-related if the combustion-related emissions are zero.

2.2.5 Historical CFC Emissions

The chlorofluorocarbons (CFCs) are a family of compounds derived from the methane (CH_4) molecule. A fully-halogenated CFC is formed by replacing all hydrogen molecules with the halogens chlorine (Cl) or fluorine (F). When the bromine (Br) atom is used as a replacement, the compounds are referred to as halons. These various species have differing effects on ozone and differing infrared absorptivities, factors that must be considered in estimating their chemical and climatic influences (see Ch. 3).

Human activities are the primary cause of the chlorofluorocarbon emissions that may

Table 2.5 Historical N_2O emissions.

Sources and sinks	Quality of emissions data	Uncertainty	Notes
Sources			
Natural sources	Poor	Great	Total global emissions from natural and anthropogenic sources are estimated to range from 10 to 19 TgN/yr. Natural sources are approximately 45–65% of global emissions. Natural sources include (in TgN/yr): oceans and estuaries, 2 ± 1; natural soils, 6.5 ± 3.5; aquifers, wildfires, lightning, and volcanos, 0.8 ± 0.3. It is not clear that all sources and sinks for N_2O have been identified. Although it is a chemically, highly stable gas in the atmosphere, the average lifetime uncertainty ranges from 100 to 175 years, leading to a 100% uncertainty in the global emission rate.
Non-energy	Poor	± 100%	35 to 55% of global emissions. Current source estimates are (in TgN/yr): natural soil cultivation, 1.5 ± 0.8; nitrogen fertilizer applications, 0.8 ± 0.4; slash-and-burn agriculture/land clearing, 0.7 ± 0.4.
Energy	Poor	Great	10 to 40% of global emissions. Fossil-fuel combustion is estimated to produce 0 to 3.2 TgN/yr. Combustion studies originally showed high rates of release of N_2O from fossil-fuel use in high-temperature combustion. Later analysis revealed a sampling artifact which, when removed, greatly reduces the emission coefficient.
Sinks			
Atmospheric chemistry	N/A	± 40%	Current sink estimates are (in TgN/yr): photolysis and reaction with $O(^1D)$ in the stratosphere, 10.5 ± 3.0; accumulation, 3.5 ± 0.5.

Source: Based on Wuebbles and Edmonds (1988).

Human activities are the primary cause of the chlorocarbon emissions that may affect climate and the distribution of stratospheric ozone ... The U.S. accounts for about one-quarter of the production of CFC-11 and about one-third of the production of CFC-12.

affect climate and the distribution of stratospheric ozone (WMO, 1985; Wuebbles and Edmonds, 1988; Watson et al., 1988; Khalil and Rasmussen, 1989a). The two most important CFCs in terms of the quantities being emitted, their ozone depletion potential, and their relative greenhouse forcing potential are CFC-11 ($CFCl_3$) and CFC-12 (CF_2Cl_2). However, while CFC-11 has a stronger ozone depletion potential (ODP) than CFC-12 (on a molecule for molecule basis), CFC-12 has more than twice the greenhouse forcing potential of CFC-11 (Dupont, 1987). Other CFCs (such as CFC-113, -114, and -115) and the halons are also strong ozone depletors (as are methyl chloroform and carbon tetrachloride), but their role in greenhouse forcing and ozone depletion are currently not as great as that of CFC-11 and CFC-12.

Total worldwide production of CFC-11 and CFC-12 is not well established. However, most CFC-producing companies in the free world report production to the Chemical Manufacturing Association (CMA, 1988). The CMA reports that the cumulative production of CFC-11 through 1987 was about 7.1 Tg while CFC-12 was about 9.4 Tg. Actual production in 1987 was 0.4 Tg/yr of CFC-11 and 0.4 Tg/yr of CFC-12. The U.S. accounts for about one-quarter of the CMA production of CFC-11 and about one-third of the production of CFC-12 (CMA, 1988).

Production does not equal emission to the atmosphere. Of the total cumulative production of 16.4 Tg of CFC-11 and -12 by 1987, 15.0 Tg (or about 90%) are thought to have been emitted to the atmosphere (CMA, 1988). The remaining CFCs are "banked" or trapped in products such as air conditioners and refrigerators (as working fluids), spray cans (as aerosols), and structural and flexible foam products (as blowing agents). The rate of emission of these banked CFCs depends upon the quantity held in relatively slow-release usages, such as home refrigerators, compared to short-term release products, such as spray cans. Nearly all aerosol uses have been banned in the U.S. since 1978. However, aerosol use of CFC-11 and CFC-12 is probably the dominant use of these two compounds outside the U.S.

2.3 Forecasts of Future Greenhouse Gas Emissions

Several things can be stated about future emissions of greenhouse gases. Scenarios of future emissions that include substantial emissions of fossil-fuel CO_2 all contain substantial reliance on coal production and use in energy conversion technologies, specifically electricity production, gasification, and liquefaction. Furthermore, the resource base of coal is dominated by concentrations of coal found in the U.S., U.S.S.R., the P.R.C., and other Organization for Economic Cooperation and Development (OECD) and Eastern European states. While land-use change can contribute substantial emissions of CO_2 continuing over several years, the cumulative potential CO_2 emissions from this source are bounded as noted in the discussion that follows. The future emission of CFCs will depend upon the behavior of participants in the Montreal protocol process. There is good prospect for declining emissions, primarily because of concerns about the stratospheric ozone depletion properties of these substances.

Forecasts of greenhouse gas emissions begin from relatively uncertain historical foundations for gases other than the CFCs and fossil-fuel CO_2. Uncertainties in the historical basis include those from estimating the absolute levels of emission from various activities, their geographic distribution, and the relative contributions of various activities. Uncertainty surrounding future emissions of all radiatively important gases is compounded by the global scale and long time horizon over which forecasts are required.

2.3.1 Forecasts of Future Fossil-Fuel CO_2 Emissions

The recoverable fossil-fuel resource base is dominated by coal resources located in great abundance in the U.S., U.S.S.R., the P.R.C., and other OECD countries. This resource base provides no practical constraint on future atmospheric CO_2 release. The atmospheric burden of carbon in 1988 totaled approximately 740 PgC. The estimated resource of fossil fuels is huge by comparison. The carbon content of conventional oil and natural gas resources is only slightly more than half as large as the current atmospheric stock of carbon. Coal resources are an order of magnitude larger than conventional oil and gas resources. The carbon content of unconventional oil resources is 55 times larger than the current atmospheric stock of carbon. The pool of carbon available for combustion might be constrained to 4000 PgC by considering only those resources recoverable under present technologies. Even this severely constrained resource definition provides no restrictive physical constraint on climate change from fossil-fuel use.

Approximately 80% of the coal resource base is thought to be in three countries: the U.S., U.S.S.R., and P.R.C. There are approximately 800 PgC in the form of coal, recoverable with today's technologies, within the jurisdictional boundaries of the world's other countries.

The recoverable [fossil-fuel] resource base . . . provides no practical constraint on future atmospheric CO_2 release.

Future-scenarios that include rapid growth in emissions of fossil-fuel CO_2 all contain substantial reliance on coal production and use in energy conversion technologies, specifically electricity production, gasification and liquefaction. Scenarios that contain greater reliance on natural gas relative to coal have lower near-term emissions trajectories than those that rely more heavily on coal in the future. Reductions in fossil-fuel CO_2 emissions by the substitution of natural gas for coal are limited by the fact that natural gas releases CO_2 as a combustion byproduct. Oil shales are presently not economically attractive despite their abundance.

Uncertainty regarding the future rate of fossil-fuel use under "business as usual" conditions remains large (Nordhaus and Yohe, 1983; Edmonds et al., 1986). The variation in emissions scenarios arises from underlying uncertainties surrounding major human activities such as economic growth, the rate of energy-efficiency improvement, and the type and rate of economic development in the developing world. The future efficiency and cost of energy conservation and renewable and nuclear technologies are also critical determinants of future emissions scenarios. Variations in the rate of population growth and the estimated resource base of fossil fuels do not produce variations in emission forecasts comparable with the previously mentioned factors. In light of their abundance, economic attractiveness, and the anticipated growth in the demands for energy services generated by global population and economic growth, fossil fuels are expected to continue to play an important role in supplying the world's energy needs.

One of many scenarios of fossil-fuel emissions is considered in Section 3.4.1.

2.3.2 Forecasts of Future CO_2 From Land-Use Change

The causes of land-use change vary from country to country, making the development of forecasts of future net emissions very difficult. The dominant factors affecting deforestation include the conversion of forests to agricultural and pasture lands, logging, and fuel wood harvests. Unlike fossil-fuel CO_2 emissions, cumulative net emissions from land-use change can be bounded from above by the total stock of terrestrial-biomass carbon. Thus, while annual emissions are highly uncertain for any given year, cumulative net emissions from the living biosphere cannot exceed about 560 PgC; if all soil carbon (humus, peat, etc.) were to be oxidized, total biogenic emissions could be a few times larger (see Chapter 3 for a discussion of the carbon budget). This compares to a current stock of carbon in the atmosphere of approximately 750 PgC.

Annual emission forecasts under business-as-usual conditions vary within a range similar to that for 1980 estimated emissions. If net reforestation occurs, net emissions due to human activities of 1 PgC/yr or less (or even carbon uptake) are possible. It is not out of the question that the terrestrial biosphere could provide a net future sink for CO_2, or even be a sink at the present time. On the other hand, some rapid deforestation scenarios contain peak emissions of 5 PgC/yr for brief periods (before exhaustion of the forest resources).

As a result of climatic warming, there exists the possibility for significant enhancement of carbon emissions via the release of carbon stored in now frozen soils and as peat and humus in other regions. Estimates of potential releases are highly speculative but potentially, within an order of magnitude,

Uncertainty regarding the future rate of fossil-fuel use remains large. The uncertainty in emissions arises from underlying uncertainties surrounding major human activities such as economic growth, the rate of energy-efficiency improvement, and the type and rate of economic development in the developing world.

equivalent to current fossil-fuel CO_2 emissions. Some fraction of this carbon could be released in the form of CO_2 or CH_4, depending on soil conditions. If CH_4 is a large fraction of the carbon released, the attendant effect on the CH_4 budget would be significantly greater than on the CO_2 budget. Another relatively short-term climatic feedback may occur from increases in the decomposition of vegetation resulting from increased mortality resulting from rapid climate change.

However, there is also the possibility that there may be a net increase in the biosphere as a result of the more moderate climate and the fertilizing effect of the CO_2 increase. For example, Gifford (1989) suggests that there may be a 30 to 40% enhancement as a result of a CO_2 doubling, although questions of the availability of nutrient and water resources deserve careful consideration.

2.3.3 Forecasts of Future Methane and Carbon Monoxide Emissions

Methane emissions are generally anticipated to contribute less than 25% to future global radiative forcing, with most estimates in the 10 to 20% range. Because the sources of CH_4 are uncertain, forecasts of emissions are also uncertain. Most forecasts simply project the rate of accumulation in the atmosphere to continue. More recently, attempts have been made to estimate future emissions on the basis of estimates of energy (coal mining, gas production, and landfills) and agricultural activities (rice cultivation and ruminant livestock production), which in turn are determined by assumptions about population, economic growth, taste, and the assumed natural-gas resource base. Studies that have attempted to develop emission forecasts from forecasts of underlying human activities, assuming business-as-usual conditions, yield increasing CH_4 emissions for the period to the year 2050, ranging from 0.5%/yr (U.S. EPA, 1989) to 1.25%/yr (Rotman et al., 1989).

Few studies have derived forecasts of future CO emissions from forecasts of underlying human activities. This is partly because of the complexity of the problem of estimating CO emissions and partly because CO is not itself a strong greenhouse gas. Carbon monoxide emission forecasts are useful only if the analysis contains an atmospheric chemistry model capable of analyzing the interactions of CO, OH, and CH_4. The U.S. EPA (1989) forecasts emissions of CO that increase in both the Rapidly Changing World (RCW) and Slowly Changing World (SCW) scenarios at approximately 0.85%/yr to the year 2050, although the time profiles of the two cases differ in details.

2.3.4 Forecasts of Future N_2O Emissions

Forecasts of future N_2O emissions vary greatly. Those constructed prior to the discovery of an artifact in the measurement technique for stack-gas emissions of N_2O link future emissions primarily to the use of coal. Emission growth rates for such studies as Rotman et al. (1989) and Mintzer (1987) for the period to the year 2050 produce rates of growth of emissions that range between 0.50 and 1.75%/yr. The U.S. EPA (1989) uses a much lower emission coefficient for fossil-fuel N_2O with a consequent lower rate of emission growth, 0.3%/yr. Most forecasts show a relatively small role for N_2O in determining future changes in infrared radiative forcing, even though an individual molecule of N_2O is estimated to be much more efficient in absorbing infrared radiation than a molecule of CO_2 (see Section 3.5) and has an extremely long average residence time in the atmosphere, 100 to 175 years. These factors are generally counterbalanced by low emissions, which increase at rates similar to those forecast for other greenhouse gases.

2.3.5 Forecasts of Future CFC Emissions

Countries representing most of the world's current production and consumption of CFCs have agreed to curtail production and use of CFCs. This agreement (the Montreal Protocol) was reached out of the desire to protect stratospheric ozone, which results indicate is being catalytically destroyed by chlorine introduced into the stratosphere by CFCs. The protocol freezes the combined production and consumption of a group of CFCs (CFCs-11, -12, -113, -114, and -115). Production and consumption limits on specific CFCs within this group may vary as long as the total ozone depletion potential (ODP) of the group (i.e., the ozone depletion potential of each molecule times its production), does not exceed that of 1986. The conditions of this agreement require production and consumption for each country to be frozen at 1986 levels as of July 1, 1989. Production and consumption are to be further curtailed to 80% of 1986 levels by July 1, 1993, and to 50% of 1986 levels by July 1, 1998.

Actual future levels of production of CFCs will depend upon the number of countries that eventually join the protocol, the amount of increase in usage in developing countries through special conditions allowed by the treaty during the first ten years, actions taken by manufacturers to change production, and any revisions to the protocol when the treaty

Given the long lifetime of CFCs, they will continue to exert a significant greenhouse effect for many decades, even if production and use decline.

is reconsidered in 1990. (This reconsideration seems likely to phase out the use of presently controlled chemicals and perhaps to restrict other ozone depleting substances such as carbon tetrachloride and methyl chloroform).

The percentage of produced CFCs that is emitted is almost certain to decline with heightened awareness of adverse environmental effects of CFC use. Many procedures that previously resulted in immediate release of CFCs will be modified, and recycling and destruction of used CFCs will become more commonplace. However, given the long lifetime of CFCs, they will continue to exert a significant greenhouse effect for many decades, even if production and use decline.

Several replacements for CFCs are under consideration. Most of them are either hydrochlorofluorocarbons (HCFCs—containing hydrogen) or hydrofluorocarbons (HFCs—hydrogen but no chlorine). These compounds have substantially shorter atmospheric lifetimes than CFCs, but a greater mass may be required to fulfill the same requirements; on balance, over the long term they will likely have a smaller radiative forcing impact. However, the ultimate impact of CFCs on greenhouse warming and on stratospheric ozone depletion depends not only on the relative ODP and infrared absorption characteristics but also on quantities produced in the future.

2.4 Research Needs

2.4.1 Historical Emissions

The state of understanding of historical emissions of greenhouse gases varies tremendously, from good for fossil-fuel CO_2 emissions and for CFCs, to fair to poor for land-use CO_2, to poor for CH_4, CO, and N_2O. To improve estimates of future concentrations, we need to improve the quality of emission estimates from known emission sources, including both natural and man-made emissions, extend estimates of those sources backward in time, and identify and quantify new emissions sources.

For CO_2 emissions, estimates of land-use change must be improved. Current estimates of the extent of deforestation are quite uncertain.

For CO_2 emissions, estimates of land-use change must be improved. Current estimates of the extent of deforestation are quite uncertain. Estimates of the biomass stock associated with land use are based on a limited number of destructive samples or official production statistics for timber and forest products combined with an expansion term. Neither emission coefficients nor estimates of the extent of deforestation are adequate.

Fossil-fuel CO_2 emissions are generally of good quality, but even here differences in estimates obtained using United Nations (U.N.) energy production and consumption figures are at odds with estimates obtained using national statistics. For example, CO_2 emissions calculated from fossil-fuel-consumption data supplied by the U.N. indicate that U.S. emissions peaked in 1979, whereas calculations based on energy-use estimates from the U.S. Department of Energy strongly suggest that the emissions in 1973 exceeded those in 1979. Since U.N. estimates are derived from data supplied by the U.S. Department of Energy, these two estimates should be identical. In general it is not possible to trace fossil-fuel CO_2 emissions to specific activities, such as passenger automobiles, for all countries of the world; and for much of the world, specifically, the U.S.S.R. and P.R.C., it is not possible to

Estimates of emissions of CH_4, CO, and N_2O remain the least certain currently available. It is not clear whether all of the important sources of emissions have been identified for these gases.

attribute emissions to end-use sectors of origin with confidence.

Estimates of emissions of CH_4, CO, and N_2O remain the least certain currently available. It is not clear whether all of the important sources of emissions have been identified for these gases. There are not even good histories of emissions for individual subcategories. In the field of energy-related emissions, better estimates of emissions could be obtained for all of the sources currently thought to be important, including natural gas production and distribution, deep coal mining, combustion, and landfills. Some progress has been made in developing estimates of CH_4 emissions from natural gas for Europe and North America. Beyond these areas, information is sparse. Specifically, emission estimates for the U.S.S.R., one of the world's largest producers of natural gas, are highly speculative. Similarly, estimates of emissions from deep-mined coal have been improved by recent research, but questions regarding emission coefficients remain, particularly with regard to long-wall mining technologies. Estimates of emissions from landfills, currently thought to be important, are crude. More sophisticated analysis and ground truth observations are needed. Combustion appears to be a minor source of CH_4 for modern technologies, but emission coefficients are not of good quality for uncontrolled combustion.

The short atmospheric lifetime and importance of local characteristics to the determination of CO emissions leave little likelihood that an exact inventory of CO emissions

A second generation of greenhouse gas emission models is needed that includes both a global integrating model and compatible national models.

will ever be obtained. Nevertheless, great improvements in the characterization of emissions are possible, especially with regard to energy source terms, although they will require a painstaking analysis of global combustion.

Improvements in N_2O emission inventories can be obtained from continued analysis of combustion processes and the ensuing atmospheric chemistry. More extensive analysis of N_2O release from soil tillage and fertilizer application is needed, including examination of associated off-site releases (e.g., resulting from the runoff of fertilizer, etc.). Natural source terms, in particular releases from natural soils, require improved emission coefficients as well as better land-use characterizations.

While CFC production figures for non-centrally planned economies are generally quite good, production figures are not yet available for the centrally planned economies. These figures should become available for the U.S.S.R. as a signatory to the Montreal Protocol. Data from nonsignatories such as the P.R.C. will remain a deficiency of the production accounting and will require improvement. The link between production and emission to the atmosphere is another weak link in the emissions inventory system that needs improvement. In addition, data are needed on emissions of other ozone-depleting species not covered by the Protocol.

2.4.2 Future Emissions

Our ability to examine scenarios of potential future greenhouse gas emissions is limited by lack of historical data on human activities and associated emission rates by country,

Multiple approaches and alternative analytical approaches to the forecast of future greenhouse gas emissions are important to provide cross-checks and insights unobtainable from a single model.

over time, against which to calibrate emission models, and by lack of modeling tools adequate to the task of integrating analyses of technologies and human activities (including energy, transportation, agriculture, manufacture, capital formation, and land use).

A second generation of greenhouse gas emission models is needed that includes both a global integrating model and compatible national models. These new models should also be capable of analyzing the emissions of all greenhouse gases from all sources, of both natural and human origin, and provide an analysis of the interplay between the variety of human and natural activities responsible for greenhouse gas emissions. It is also important that these new models be developed in conjunction with related scientific efforts in the areas of carbon-cycle atmospheric chemistry and climate change. Estimating natural sources accurately will require that the model be connected to a carbon cycle model (see Chapter 3) in that carbon emission and uptake rates are dependent on the CO_2 concentration and other factors.

The first-generation models focused on a particular aspect of the emission problem without regard to how that activity interacted with other human and natural activities. The Edmonds-Reilly model (Edmonds and Reilly, 1985), for example, is a global energy model in which the production of commercial biomass has no interaction with the agricultural sector or land use in general. Most radiative forcing forecasts have also been made without regard to the interaction of emissions and atmospheric processes. Those that attempt to model this interaction explicitly, for example U.S. EPA (1989) and Rotman et al. (1989), produce emission estimates using a set of parallel but unrelated models. This approach has been successful in mapping out a range of uncertainty, assuming future emissions under business-as-usual circumstances and in providing preliminary analyses of the potential for emissions reduction and associated reductions in atmospheric concentrations of greenhouse gases. But this approach is inadequate to the task of analyzing the link between technical options and the cost of reducing emissions, and similarly inadequate for understanding how activities in one arena of human or natural system activity interact with those in another.

The link between technology and emissions has been studied using detailed technology analysis (e.g., Goldemberg et al., 1987). Although this approach is generally very rich in technological detail, it is weak in its description of the relationship between technologies and the overall, "macro," scale of human activities, for example, aggregate energy production and use, and Gross National Product. There is a need for new models capable of linking the so-called "bottom up" approach of researchers such as Goldemberg et al. and the so-called "top down" approach of researchers such as Edmonds and Reilly.

Other important avenues of research are the development of improved characterizations of technological options for reducing greenhouse gas emissions and the development of country-specific expertise for key greenhouse gas emitters, specifically such countries as the U.S., U.S.S.R., P.R.C., Japan, the United Kingdom, India, Brazil, the Federal Republic of Germany, Mexico, Canada, Indonesia, Australia, and eastern European countries including Czechoslovokia, Poland, Hungary, and the German Democratic Republic.

Although developing a second-generation model is an important research priority, so too is the development of a set of alternative analysis techniques. Multiple approaches and alternative analytical approaches to the forecast of future greenhouse gas emissions are important to provide cross-checks and insights unobtainable from a single model.

Chapter 3: THE CHANGING COMPOSITION, CHEMISTRY, AND RADIATIVE FORCING OF THE GLOBAL ATMOSPHERE

3.1 The Changing Atmospheric Composition

Atmospheric trace gas concentrations are changing. Concentrations of CO_2, CH_4, N_2O, CO, and several chlorinated and brominated halocarbons (especially the chlorofluorocarbons) are known to be increasing (see Table 3.1). Stratospheric concentrations of ozone are decreasing while tropospheric concentrations appear to be increasing. These changes are thought to result primarily from human activities.

Many of these gases can directly intensify the greenhouse effect through their absorption of infrared radiation. The increasing concentration of CO_2, largely as a result of fossil-fuel burning, has received the most attention. As is discussed later, however, climate models indicate that the sum of the radiative effects from the growing atmospheric concentrations of the other gases, along with corresponding effects on the atmospheric distribution of ozone, could be as large as the climatic impact projected for CO_2 alone. Unlike the overall extent of climate change, the direct radiative forcing on the atmosphere resulting from the increasing concentrations of CO_2 and other greenhouse gases can be calculated reasonably well for clear sky conditions, although uncertainties of about 10% still remain. (The effects of changes in cloud cover significantly amplify this uncertainty; see Chapter 4.)

In addition to direct radiative effects, many of these gases have indirect radiative effects on climate through atmospheric chemistry. Both measurements and theoretical models indicate that the distribution of ozone in the troposphere and stratosphere is changing as a result of such interactions. Because

Concentrations of CO_2, CH_4, N_2O, CO, and several chlorinated and brominated halocarbons (especially the chlorofluorocarbons) are known to be increasing.

of vertical variations in ozone (O_3) concentration and temperature, O_3 is not only a greenhouse gas (primarily in the troposphere) but also the primary absorber of ultraviolet radiation at wavelengths from about 200 to 300 nm (primarily in the stratosphere). Concern about changes in the amount of ultraviolet radiation reaching the Earth has brought about much of the attention to ozone, but its contribution to climate change is also important.

There are other concerns regarding chemical interactions in the atmosphere. The increasing concentration of CH_4 is expected to cause an increase in the stratospheric concentration of water vapor as a result of the chemical destruction of CH_4 (each 1 ppmv increase in the CH_4 concentration leads to an increase in stratospheric H_2O of about 2 ppmv). The added water vapor results in additional climatic warming. Many of the greenhouse gases of interest react with OH in the troposphere; increasing emissions of CH_4 and CO can reduce OH concentrations, resulting in longer lifetimes for these and other greenhouse gases of concern. Even CO_2, which has no known chemical interactions in the atmosphere, affects levels of stratospheric ozone through its radiative effect, which reduces stratospheric temperatures and results in a slowing down of ozone catalytic destruction and a net increase in stratospheric ozone.

This chapter examines the known changes occurring in the distributions of many of

Table 3.1 Concentrations and lifetimes of important atmospheric trace gas constituents.

Gas	Nonurban tropospheric concentration[a] (ppmv)	Trend in atmospheric concentration[a] (% per year)	Atmospheric lifetime[b] (years)	Primary removal processes
CO_2	351 (1988)	~0.4%	~250[c]	Uptake by the oceans.
CH_4	1.7	~1	~10	Chemical transformation to CO_2 and H_2O.
CO	0.12(N.H.) 0.06(S.H.)	~1(N.H.) ~0(S.H.)	~0.3	Chemical reaction to CO_2.
N_2O	0.31	~0.3	~150	
NO_x (=NO + NO_2)	$1–20 \times 10^{-5}$	unknown	≤ 0.02	Chemical reaction to nitric acid and removal by precipitation.
NO_y(=NO_x + HNO_3 + PAN + N_2O_5)	$4–20 \times 10^{-4}$	unknown	≤ 0.02	
$CFCl_3$	2.6×10^{-4}	~5	~70	Photodissociation to chlorine and hydrochloric acid that is subsequently removed by precipitation.
CF_2Cl_3	4.4×10^{-5}	~5	~120	
$C_2Cl_3F_3$	3.2×10^{-5}	~10	~90	
CH_3CCl_3	1.2×10^{-4}	~4.5	~6	
CF_2ClBr	1×10^{-6}	~12	~12–15	Dissociation and removal by precipitation.
CF_3Br	1×10^{-6}	~15	~110	

[a] Corresponds to mid-1980s values.

[b] The estimates of atmospheric lifetimes have an uncertainty of about 10–20% for many of these species. A mid-range value is given.

[c] Computed as the ratio of the atmospheric burden to net annual removal. Net annual removal is estimated as emissions less atmospheric accumulation.

Table 3.1 Concentrations and lifetimes of important atmospheric trace gas constituents (continued).

Gas	Nonurban tropospheric concentration[a] (ppmv)	Trend in atmospheric concentration[a] (% per year)	Atmospheric lifetime[b] (years)	Primary removal processes
SO_2	$1\text{–}20 \times 10^{-5}$	unknown	~ 0.02	Dry deposition and conversion to sulfate that is removed by precipitation.
COS	5×10^{-4}	<3	2–2.5	Conversion to sulfate and removal by precipitation.
DMS	$5\text{–}20 \times 10^{-5}$	unknown	~ 0.01	Chemical reaction leading to removal by precipitation.
NMHC	<1	unknown	≤ 0.02	Reaction with O_3 and OH.
O_3	0.02–0.1	0.5–1	<0.1	Chemical reaction or removal at surface.
OH	$4\text{–}100 \times 10^{-9}$	unknown	<0.01	Chemical reactions to H_2O.
H_2O	$1\text{–}20 \times 10^{3}$	unknown	~ 0.03	Precipitation or chemical reaction.

[a] Corresponds to mid-1980s values.
[b] The estimates of atmospheric lifetimes have an uncertainty of about 10–20% for many of these species. A mid-range value is given.

these constituents and the potential for the changes in radiative forcing and chemical interactions that directly and indirectly influence climate and ozone.

3.2 Carbon Dioxide: Observations and Trends

3.2.1 Observed Changes in Atmospheric CO_2 Concentration

Continuous measurements of CO_2 at Mauna Loa Observatory, Hawaii, begun in 1958 by Keeling and co-workers (Keeling et al., 1976) and continuing also as part of the National Oceanic and Atmospheric Administration monitoring program (Komhyr et al., 1985), provide the best record of the change in concentration of this greenhouse gas. The CO_2 concentration is now measured routinely at a global network of stations carefully located to avoid local sources that might bias the observations. Continuous measurements are made by the U.S. at Mauna Loa, Hawaii; Barrow, Alaska; American Samoa; and the South Pole. Flask samples are collected at 20 additional sites, as shown in Figure 3.1.

Data collected by this sampling program reveal spatial patterns in CO_2 variations that show the influences of exchanges between the atmosphere and other reservoirs. Interestingly, the Mauna Loa record, which is the longest such record available, appears to provide a reasonable estimate of the global mean concentration.

Monthly and annual average CO_2 concentrations at Mauna Loa since 1959 are plotted in Figure 3.2; the 1988 concentration was about 351 ppmv. Between 1959 and 1988, the annual average CO_2 concentration at Mauna Loa increased from 316 ppmv to 351 ppmv, or about 11%. The annual cycling, which is this strong only in the Northern Hemisphere, is produced by the net uptake of carbon during the growing season that results from increased photosynthesis and the net release during the dormant season that results from domination by respiration. This net seasonal exchange in the Northern Hemisphere amounts to about 7 PgC. There are some indications that the amplitude of the annual cycling is increasing, which may indicate that the biosphere is growing (at least in middle latitudes), but this is quite uncertain and its effect on carbon storage is not established.

The changing CO_2 concentration in air bubbles trapped in glacial ice indicate changes in atmospheric levels prior to the start of direct measurements in 1958. Data obtained from an ice core extracted at Siple Station, Antarctica (Neftel et al., 1985; Friedli et al., 1986) show the history of CO_2 from the middle of the 18th century, prior to the onset of significant fossil-fuel use, to the time of the Mauna Loa record, Figure 3.3. The ice core record, which agrees well with modern measurements, indicates an 18th century CO_2 concentration of about 278 ppmv; the 1988 concentration of 351 ppmv represents a 26% increase.

3.2.2 The Global Carbon Cycle

Interactions between the atmosphere and other reservoirs that form the global carbon cycle determine the change in atmospheric CO_2 concentration that occurs with the addition of CO_2 from fossil fuels (see Figure 3.4). The increase in concentration from 1959 to 1988 at Mauna Loa corresponds to an increase in the carbon content of the atmosphere of about 75 PgC. During this same period, about 135 PgC were released to the atmosphere from fossil fuels and by the manufacture of cement (extended from Marland et al., 1989). The change in the carbon content of the atmosphere is less than the fossil-fuel release—the ratio is about 0.56—because of carbon exchanges between the atmosphere and other reservoirs.

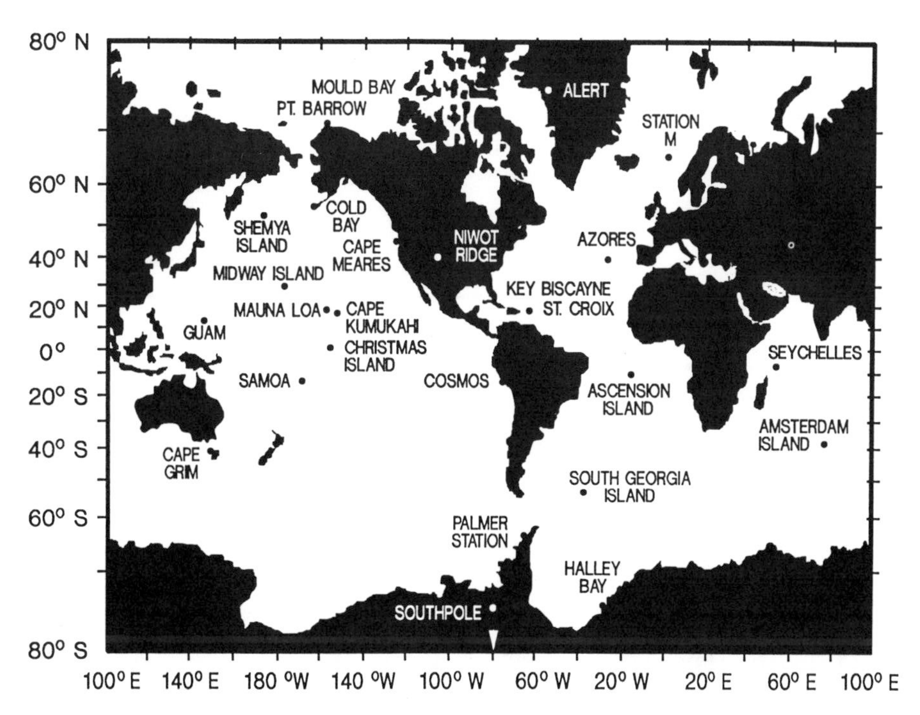

Figure 3.1 Locations of measurement stations and sampling sites for collection of data on the atmospheric CO_2 concentration.

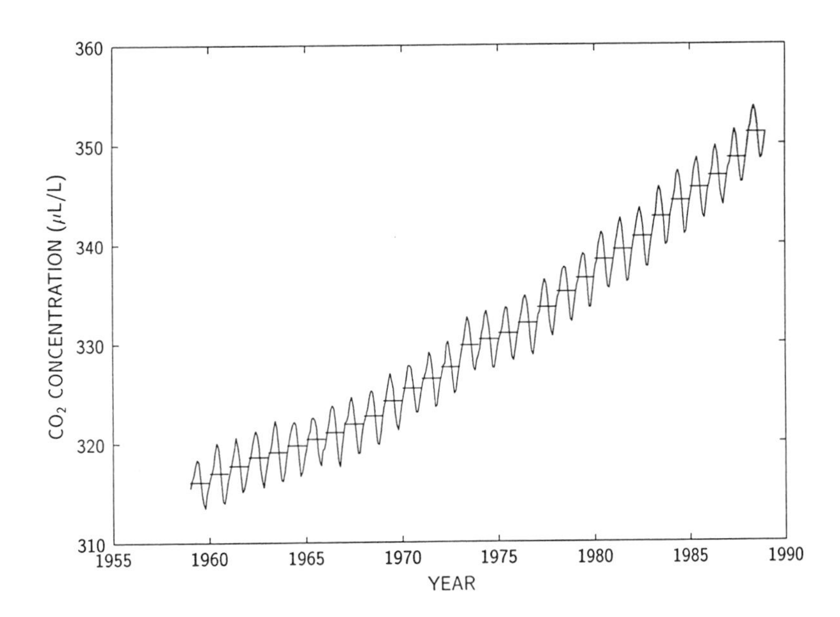

Figure 3.2 Atmospheric CO_2 concentration at Mauna Loa Observatory, Hawaii (Keeling 1986). Lines connect monthly average values to form the oscillating curve. Horizontal bars indicate annual average concentrations.

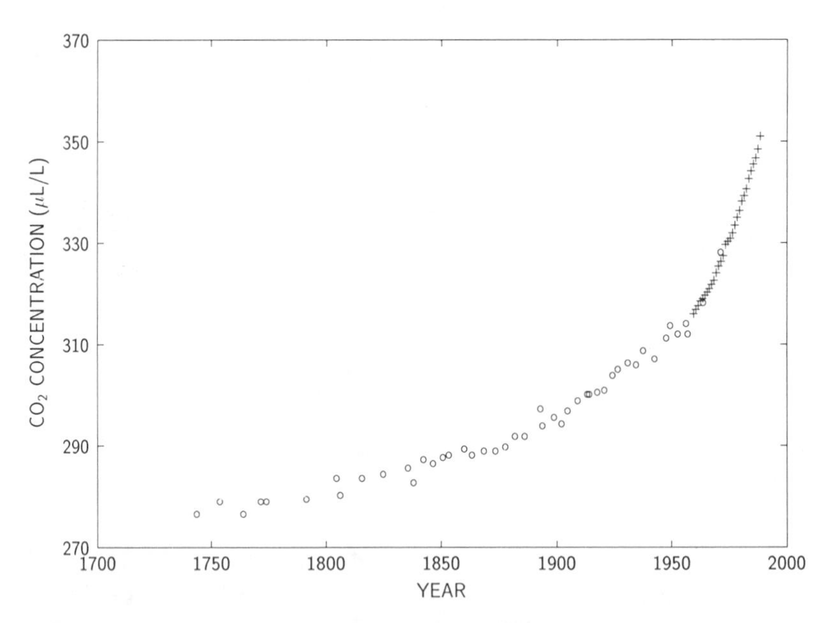

Figure 3.3 Concentration of CO_2 in air bubbles trapped in an ice core extracted at Siple Station, Antarctica (o) (Friedli et al., 1986; Neftel et al., 1985) and as recorded at Mauna Loa since 1959 (+).

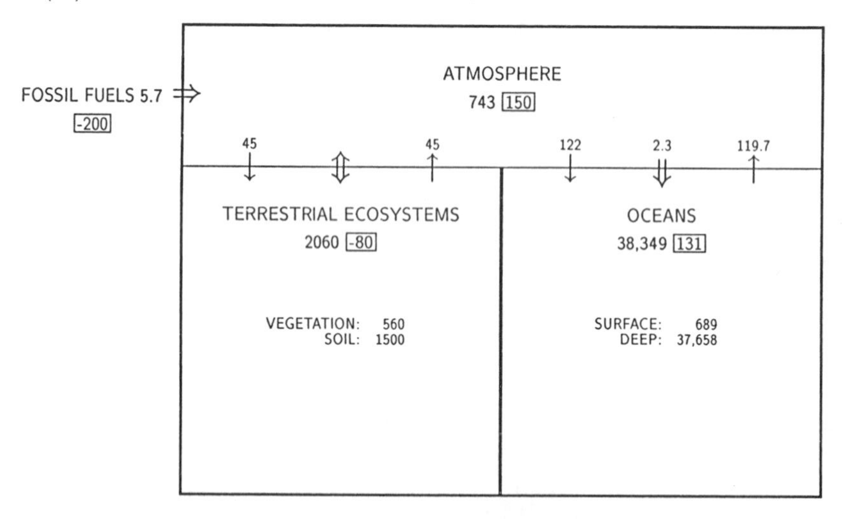

Figure 3.4 The major reservoirs and fluxes of the global carbon cycle that determine the response of atmospheric CO_2 concentration to fossil-fuel releases (Killough and Emanuel, 1981; Emanuel et al., 1984). The oceans are the primary sink for carbon from the atmosphere. Vegetation and soil may contribute carbon to the atmosphere in addition to that from fossil fuels because of disturbance by human activities, or under other circumstances, serve as a carbon sink in addition to the oceans. Reservoir contents and fluxes approximate 1987 values. Boxed values are cumulative changes in reservoir contents between 1745 and 1987 in a model solution with the atmosphere constrained by the Siple ice-core and Mauna Loa CO_2 measurement records.

The ice core record, which agrees well with modern measurements, indicates an 18th century CO_2 concentration of about 278 ppmv; the 1988 concentration of 351 ppmv represents a 26% increase.

The oceans, which currently contain about 50 times more carbon than the atmosphere, are the most important sink for carbon from the atmosphere. Carbon can also be sequestered on land in vegetation and soil, but the turnover is more rapid than for the oceans. In many regions, land-use changes, particularly forest clearing, lead to net CO_2 releases in addition to those from fossil fuels.

The rate of CO_2 uptake by the oceans is determined by three sets of processes: (1) CO_2 exchange across the air-sea boundary; (2) the incorporation of surface water carbon into compounds other than CO_2; and (3) mixing and circulation, which transport carbon from surface water into deeper layers where it can be sequestered from atmospheric exchange.

Plants assimilate atmospheric CO_2 by photosynthesis; CO_2 is released by the decomposition of dead organic matter and by fire. For reasonably undisturbed ecosystems, uptake and loss fluxes are frequently assumed to balance over sufficiently long time periods. The balance certainly shifts on shorter time scales with changes in nutrient availability, climate, and sporadic natural disturbances.

An increasing atmospheric CO_2 concentration may stimulate plant productivity, making vegetation and soils more effective in sequestering carbon from the atmosphere. But in many regions, forests are being harvested or cleared and replaced by ecosystems having substantially less carbon storage. Such land-use changes have led to additional net CO_2 releases to the atmosphere. Prior to human intervention, the amount of carbon stored in vegetation is estimated to have been about 1000 PgC, nearly double the estimate of current content (Trabalka, 1985).

In addition to the major carbon exchanges between the atmosphere and the oceans and between the atmosphere and vegetation and soil that appear to dominate the response of the atmospheric CO_2 concentration to fossil-fuel releases, numerous other carbon fluxes may also be significant. Furthermore, atmospheric exchanges and the turnover of carbon in the oceanic and terrestrial components of the carbon cycle depend on climate so that climate change caused by increases in greenhouse gas concentrations may very well feed back on the processes controlling atmospheric composition and the global carbon cycle.

For example, simulations of biogeochemical feedbacks (e.g., ocean CO_2 uptake, methane release, changes in albedo of global vegetation) have indicated that such processes could, in combination, substantially increase climate change (Lashof, 1989). Also, arctic and boreal wetlands of the tundra are known to be important sources of methane (cited in Mooney et al., 1987) and it has been suggested that global warming might accelerate the release of methane (assuming no significant change in arctic vegetation) and further contribute to climate change (National Research Council, 1988). Thus, understanding of terrestrial ecosystems (Mooney et al., 1987) and the oceans is crucial to our ability to predict regional- and global-scale changes in atmospheric chemistry and climate.

3.2.3 Carbon Cycle Models

Observational studies of the phenomena controlling the CO_2 concentration are often impractical because of the time scales involved (decades to centuries), the vast extent and spatial heterogeneity of biospheric systems, and the small perturbations to natural levels that fossil-fuel releases cause in very large reservoirs such as the oceans. Furthermore, many mechanistically important variables cannot be measured directly. Mathematical models are an important means of contending with these limitations of direct studies. They also provide a means of synthesizing the diverse data assembled to address the CO_2 issue and of interpreting these data in terms of changes in atmospheric CO_2 levels.

An accounting of the oceanic uptake of carbon from the atmosphere is the most important requirement for explaining past changes in the CO_2 concentration and for estimating the magnitude of the future CO_2 increases that are expected to result from different levels of fossil-fuel use. Past changes in terrestrial carbon storage, owing to both natural phenomena and human activities, will also be important, unless the magnitude of fossil-fuel releases increases substantially. Reconciling changes in terrestrial carbon storage is crucial to understanding past changes in the atmospheric CO_2 level and in the rate of future oceanic uptake of carbon.

Oeschger et al. (1975) proposed a model of carbon turnover in the atmosphere and oceans that has become a benchmark against which other models are referenced and compared. In the Oeschger model, the vertical transport of carbon in a globally averaged water column is represented by a diffusion equation with constant diffusivity. The diffusivity parameter and the invasion flux of carbon from the atmosphere are set to achieve agreement between the equilibrium distribution of radiocarbon implied by the model and an idealized profile derived from observations.

The effects of surface-water chemistry on oceanic uptake of CO_2 can be accounted for by solving the equilibrium conditions for the significant reactions. In this approach the partial pressure of dissolved CO_2 in surface waters is a nonlinear function of their total inorganic carbon content. A linear approximation of this relationship is used in many models.

Interactions between the terrestrial and oceanic reservoirs are weak except through the atmosphere. If the exchanges between the atmosphere and terrestrial systems are assumed to be independent of the atmospheric CO_2 concentration (this assumption is made in many modeling studies that concentrate on land-use releases from vegetation and soil), then models of the atmosphere-ocean system can be decoupled from those that account for changes in terrestrial storage. Terrestrial sources can then be introduced as net fluxes into the atmospheric compartment of atmosphere-ocean models in a manner similar to the fossil-fuel CO_2 input. Stand-alone models can be used to simulate the time course of the net terrestrial source, given the intensity of forest harvest and changes in the areal extents of different ecosystem types such as conversion from forest to cropland.

The history of net CO_2 releases from terrestrial pools is uncertain. Houghton et al. (1983) developed three reconstructions of changes in terrestrial carbon storage based on land-use and demographic data with synoptic functions specifying variations in carbon storage following disturbance in different ecosystems. The estimates of historical terrestrial carbon releases are inconsistent with the Siple ice core, the Mauna Loa CO_2 records, and estimates of oceanic uptake derived from models of carbon turnover in the atmosphere and oceans (Enting and Mansbridge, 1987). As a result, projections of future concentrations are highly uncertain.

Until these inconsistencies are resolved, one recourse is to match the history of atmospheric CO_2 concentration to the Siple ice core and Mauna Loa records by calculating the net residual flux into or out of the atmosphere required to balance simulated oceanic

The estimates of historical terrestrial carbon releases are inconsistent with the Siple ice core, the Mauna Loa CO_2 records, and estimates of oceanic uptake derived from models of carbon turnover in the atmosphere and oceans. As a result, projections of future concentrations are highly uncertain.

uptake. This procedure takes the model solution from initial equilibrium conditions to the present in a way that is consistent with our best understanding of past changes in atmospheric CO_2 concentration. Future changes can then be projected if we assume inputs of CO_2 into the atmosphere from fossil fuels and land-use changes.

3.3 Other Atmospheric Constituents

This section describes the distributions and concentration trends of the non-CO_2 greenhouse gases, with emphasis on those gases whose concentration trends suggest a potentially significant effect on the state of future climate. Aerosols and other atmospheric constituents may also influence climate, and their potential influence is discussed.

3.3.1 Methane

Although the atmospheric abundance of CH_4 is less than 1/200 that of CO_2, it is an important greenhouse gas that can have a significant impact on future climate. The globally averaged atmospheric concentration of CH_4 is currently about 1700 ppbv (Figure 3.5) and is increasing by about 16 ppbv, about 1%, per year (Blake and Rowland, 1988; Khalil and Rasmussen, 1987). Global measurements indicate that CH_4 concentrations are highest at latitudes poleward of 30°N. That the CH_4 concentration in the Northern Hemisphere is about 140 ppbv greater than in the Southern Hemisphere is not surprising given that the dominant sources of methane are natural and anthropogenic sources on land. Major uncertainties exist regarding the cause (or causes) of the increase in the methane concentration and the likely future growth rate.

Destruction of atmospheric methane occurs primarily through reaction with hydroxyl (OH). Reduction in the amount of global atmospheric OH during recent decades, which may be occurring as a result of emissions of CO and volatile organic components, may provide an explanation for as much as 20 to 50% of the CH_4 increase, but this is highly uncertain, given the paucity of measurements of the tropospheric OH concentration. Methane chemistry can also affect ozone concentrations in the troposphere and stratosphere, and methane destruction increases the stratospheric water vapor concentration as its concentration increases. Atmospheric measurements indicate that the lifetime of CH_4 in the global atmosphere is about ten years.

Ice core data going back 160,000 years indicate that the concentration of methane in the preindustrial atmosphere was less than half the present concentration (Pearman et al., 1986; Khalil and Rasmussen, 1987; Raynaud et al., 1988). The methane concentration was about 700 ppbv until the beginning of the 19th century (Figure 3.6). Concentrations during glacial periods were even smaller, being as low as 350 ppbv (Raynaud et al., 1988).

3.3.2 Chlorofluorocarbons

The chlorocarbons receiving the most attention, primarily because of their larger concentrations and potentially significant effects on stratospheric ozone, are the chlorofluorocarbons $CFCl_3$ (referred to as CFC-11) and CF_2Cl_2 (CFC-12). Other similar compounds, including CCl_4, CH_3CCl_3, and the halons are currently exerting less, but still significant,

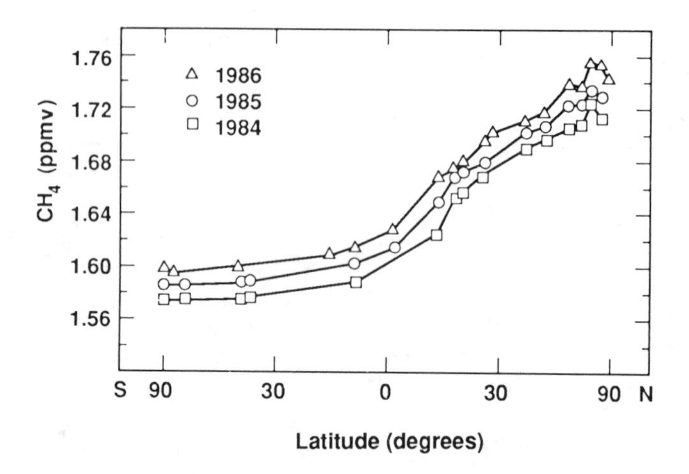

Figure 3.5 Measured methane concentrations with latitude from 1984 to 1986 based on data from the NOAA Geophysical Monitoring for Climate Change network (J. Peterson, private communication, 1988).

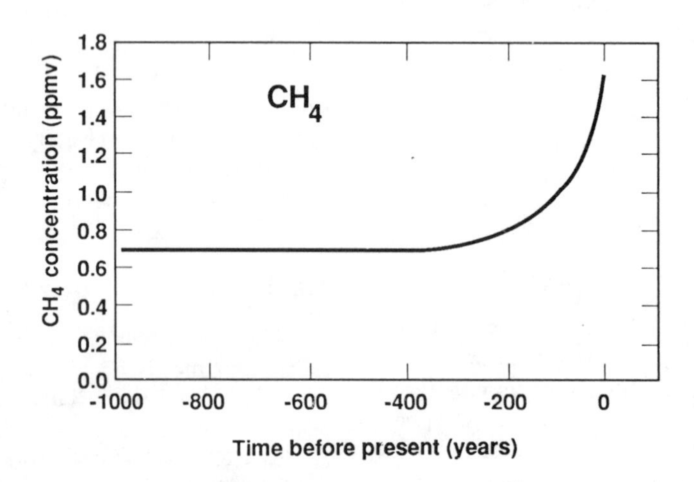

Figure 3.6 Ice core and atmospheric measurements of methane for the last 1000 years (based on Khalil and Rasmussen, 1987).

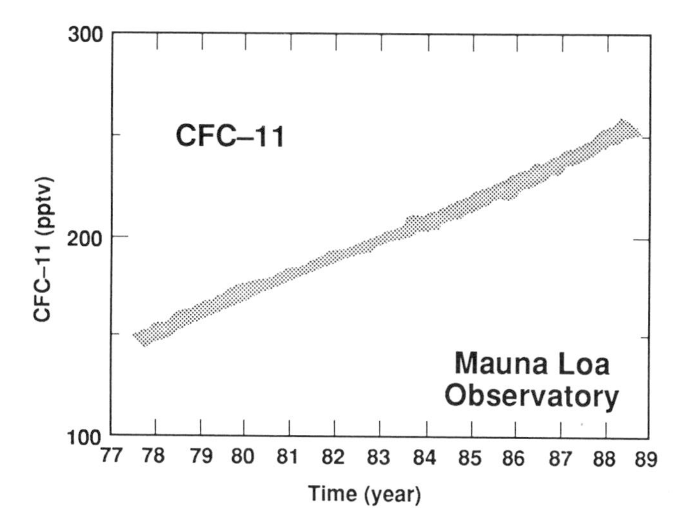

Figure 3.7 Measurements of CFC-11 at Mauna Loa Observatory since mid-1977, based on NOAA GMCC data (J. Peterson, private communication, 1989). The band represents the spread of daily and seasonal variations in concentration.

influence; their effects could become quite large if present emission trends continue.

Chlorofluorocarbons CFC-11 and CFC-12 have the highest concentrations of the man-made chlorocarbons, 0.26 and 0.44 ppbv, respectively, in 1989. The surface air concentrations of these two gases are currently increasing at a rate of over 4% per year (Figure 3.7). CFC-11 is used primarily as a blowing agent for plastic foams and in aerosol cans (other than in the U.S.), while CFC-12 is used primarily as a refrigerant and in aerosol cans.

Other important chlorocarbons include CFC-113 ($CF_2ClCFCl_2$), CFC-22 (CHF_2Cl), and methyl chloroform (CH_3CCl_3). The atmospheric concentration of CFC-113 is increasing by about 11% per year, with the present surface air concentration being about 0.03 ppbv. Abundances of CFC-22 and CH_3CCl_3 are 0.09 and 0.14 ppbv; concentrations are increasing at rates of about 7 and 4.5% per year, respectively. The species CFC-113 and CH_3CCl_3 are used primarily as solvents, while CFC-22 is used in air conditioning and refrigeration equipment. Carbon tetrachloride is used primarily as a feedstock in the production of CFCs by the chemical industry. Halons are used extensively in fire extinguishing applications.

All of these synthesized chlorocarbons have relatively long atmospheric lifetimes (i.e., relatively slow removal rates). Methyl chloroform has the shortest chemical lifetime, about six to seven years, while CFC-12 has the longest lifetime, about 120 years. The long lifetimes contribute to the rapidly increasing concentrations of these gases. The fully halogenated chlorocarbons, particularly CFC-11 and CFC-12, are of primary concern in affecting stratospheric ozone concentrations because they are destroyed in the stratosphere, primarily by photolysis, which releases all of their chlorine atoms to act as catalysts for ozone destruction. Other chlorinated halocarbons can also release chlorine to the stratosphere, whereas the destruction of halons releases bromine. The chlorine and bromine atoms can then react catalytically to destroy ozone.

Because many applications of these chemicals provide important societal benefits, there has been a scramble to find replacements, particularly for refrigeration, air conditioning, and foam blowing. Replacement compounds being suggested as a result of the Montreal Protocol are mostly chlorocarbons containing hydrogen, which will react with OH in the troposphere. The suggested replacement compounds, such as HCFC-123, HCFC-141b, and HFC-134a, are generally halogenated hydrocarbons, some containing chlorine (the HCFCs) and others not (the HFCs). All of the suggested replacements have shorter lifetimes, generally less than 20 years, and their concentrations will, as a result, be lower than the concentrations of the compounds they are replacing. Nonetheless, most of these compounds are greenhouse gases and could affect climate if concentrations become large enough.

3.3.3 Nitrous Oxide

The atmospheric concentration of nitrous oxide is increasing at a rate of about 0.3% per year (Watson et al., 1988; Khalil and Rasmussen, 1988b). The current tropospheric concentration is about 307 ppbv. Ice core data indicate that the concentration of N_2O from 3000 to 150 years before the present was about 285 ppbv (Pearman et al., 1986; Khalil and Rasmussen, 1988b). Considerable uncertainties remain in the budget for N_2O, particularly regarding the source terms and the causes of the increasing concentration.

The primary sink for nitrous oxide is photolysis in the stratosphere; it is not chemically reactive in the troposphere. Its lifetime is about 120 to 150 years. Reaction of N_2O with excited oxygen atoms is the primary source of stratospheric nitrogen oxides, which catalytically react with stratospheric ozone. Thus, an increasing N_2O concentration could lead to a net destruction of stratospheric ozone.

3.3.4 Ozone

Unlike the other trace constituents considered, ozone is not emitted by human activities. In addition, its vertical distribution is of major importance in determining its impact on climate. About 90% of atmospheric ozone is present in the stratosphere (approximately 10 to 50 km above the Earth's surface), with most of the rest in the troposphere (the first 10 km above the surface). Ozone is produced by natural processes in the stratosphere as a consequence of chemical reactions initiated by the photolysis of oxygen molecules. Destruction results primarily from catalytic reactions with chlorine oxides, nitrogen oxides, and hydrogen oxides. Although concentrations of these species are much smaller than that of ozone, a series of catalytic reactions allow a single molecule to effectively destroy a thousand or more molecules of ozone. Tropospheric ozone comes from downward transport of stratospheric ozone and from local photochemical production promoted by surface emission of nitrogen oxides and nonmethane hydrocarbons.

Measured trends of changes in ozone from ground-based measurements and satellite data indicate that ozone concentrations in the upper stratosphere are decreasing and that overall total ozone amounts at northern midlatitudes have decreased since 1969 (Watson et al., 1988; DeLuisi et al., 1989; WMO, 1989). The observed upper stratospheric decrease in ozone is comparable to the decrease projected (Wuebbles and Kinnison, 1988; DeLuisi et al., 1989) over this time period from the effects of CFCs, other trace gases, and solar flux variability (i.e., the 11-year solar cycle). The observed decrease in total column ozone was 1.7 to 3.0%

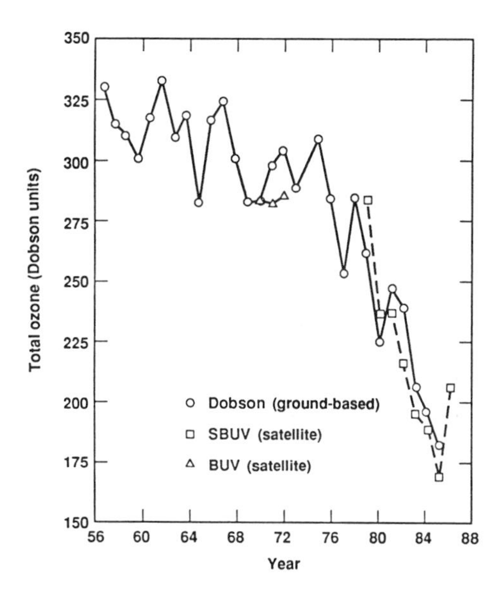

Figure 3.8 Observed decrease in springtime total ozone over the Halley Bay Station in Antarctica (Solomon, 1988; Watson et al., 1988).

over this period at latitudes between 30 and 64°N, with the largest decrease occurring in the winter months at the higher latitudes. Tropospheric concentrations of ozone appear to have increased (Logan, 1985; Tiao et al., 1986; WMO, 1989) in a pattern consistent with human-related emissions of nitrogen oxides and hydrocarbons, although uncertainties are too large to allow the establishment of a global trend.

In the Southern Hemisphere, a large decrease in the abundance of springtime Antarctic ozone has been noted over the last decade (Figure 3.8). This sudden decrease has been termed the Antarctic "ozone hole." Ozone decreases of more than 50% in the total column as compared to historical values have been observed. Measurements made in 1987 show that more than 95% of the ozone in the region from 15 to 20 km had disappeared during the September-October period. Observations also indicate that the unique meteorology during the winter and spring over Antarctica sets up special conditions forming an isolated air mass (polar vortex). Polar stratospheric clouds form if the temperatures are cold enough within this vortex region, as is often the case near the Antarctic tropopause. Chemical reactions on the cloud particles allow chlorine to be in a very reactive state with respect to ozone once the Sun appears in early spring. The weight of the evidence strongly indicates that man-made chlorine species (Cl produced from the dissociation of CFC-11, CFC-12, etc.) are primarily responsible for the observed ozone decrease within the vortex.

Although concentrations of chlorine oxides, nitrogen oxides, and hydrogen oxides are much smaller than that of ozone, a series of catalytic reactions allow a single molecule to effectively destroy a thousand or more molecules of ozone.

Because north polar temperatures in the lower stratosphere are not generally as cold as over Antarctica, and because the vortex in the Northern Hemisphere tends to break down earlier, a springtime ozone decrease in the Arctic has not yet been measured. Measurements made in the winter of 1988–89 suggest such a decrease could be possible in the future; high ClO concentrations, necessary for major O_3 destruction, were found in the vortex (WMO, 1989). A highly speculative hypothesis suggests that the greenhouse effect from CO_2 and other gases, which causes a cooling of stratospheric temperatures, could further intensify the Antarctic "ozone hole" and lead to an analogous springtime hole in the northern polar region.

3.3.5 Stratospheric Water Vapor

The increase in the water vapor concentration in the troposphere will have an important amplifying influence on the direct radiative influences of the anthropogenic greenhouse gases. Tropospheric water vapor concentrations will increase with increasing tropospheric temperatures. The effect of an increased amount of tropospheric water on stratospheric concentrations is unknown because of uncertainties in the mechanism determining the transport of tropospheric water vapor to the stratosphere. However, changes in the concentration of stratospheric water vapor can also affect climate. Such changes could occur as a result of chemical interactions in the stratosphere. Because of a poorly understood freeze-drying process at the tropopause, very little tropospheric water vapor normally penetrates into the stratosphere. The lower stratosphere is relatively dry (3 to 4 ppmv) compared to the troposphere. Concentrations of water vapor increase with altitude, to about 6 ppmv in the upper stratosphere, owing to production of H_2O by the oxidation of methane. Thus, stratospheric water vapor concentrations should increase as methane increases. Because water is one of the primary absorbers of infrared radiation, along with CO_2 and O_3, an increase in water would further add to the greenhouse effect of the increasing methane (Wuebbles et al., 1989). Changes in the mechanisms determining lower stratospheric water vapor could further affect climate.

3.3.6 Other Atmospheric Constituents

The hydroxyl radical, OH, is not itself a greenhouse gas, but it is extremely important as a chemical scavenger of many trace gases in the troposphere. Hydroxyl is the primary tropospheric scavenger of CH_4, CO, CH_3CCl_3, CH_3Cl, CH_3Br, H_2S, SO_2, DMS (dimethylsulfide), and other hydrocarbons and hydrogen-containing halocarbons. The concentration of OH affects the atmospheric lifetime of these gases and therefore contributes to their impact on ozone and climate. Reactions with OH in the troposphere limit the amount of CH_4 and halocarbons reaching the stratosphere, where these species can lead to changes in the ozone distribution. In the troposphere, OH and other related hydrogen oxides play a central role in the production of ozone. The hydroxyl radical is formed in the troposphere through the photolysis of ozone and subsequent reaction of the excited oxygen atom with water vapor. The primary removal of OH is through reactions with CH_4 and CO. The global distribution of OH has not been measured (Prinn et al., 1987; WMO, 1989).

Carbon monoxide is also not a greenhouse gas; however, it is extremely important because of its reactions with the hydroxyl radical, OH. Peak concentrations of about 200 ppbv occur at high northern latitudes, with minimum concentrations of about 50 ppbv found throughout the Southern Hemisphere. Concentrations are increasing by about 1.1%

If the CO concentration continues to increase, the resulting decrease in the OH concentration could increase the lifetime (and produce larger concentrations) of important greenhouse gases such as CH_4.

per year globally, although there is little evidence of an increase occurring in the Southern Hemisphere (Khalil and Rasmussen, 1988a; Cicerone, 1988). Carbon monoxide has a relatively short lifetime, on the order of a few months, before being transformed to CO_2; because of the high efficiency of combustion processes, this source of CO_2 is relatively minor. Nearly half of the emissions of CO are from human-related sources (40 $\pm$ 20%). As a result, its measured concentrations over land are often affected by local sources, leading to difficulty in establishing long-term trends. Nonetheless, if the CO concentration continues to increase, the resulting decrease in the OH concentration could increase the lifetime (and produce larger concentrations) of important greenhouse gases such as CH_4.

Emissions of nitrogen oxides from combustion sources have likely increased over recent decades, resulting in increased tropospheric concentrations over continents, in the boundary layer especially, and in commercial aircraft flight corridors. Nitrogen oxide emissions in the troposphere can influence local ozone concentrations and may play an important role in the apparent global increase of tropospheric ozone. Nitrogen dioxide is an important absorber of visible solar radiation and could affect climate directly if concentrations become large enough.

Several of the nonmethane hydrocarbons (NMHC) are greenhouse gases, but it is unlikely, because of their short atmospheric lifetimes, that their atmospheric concentrations will become large enough to have a major direct effect on climate, at least within the next several decades. However, NMHC emissions do affect tropospheric chemistry and could influence future amounts of O_3 and OH.

3.3.7 Aerosols

An increase in atmospheric aerosols could change climate, although in a manner different from the effect of CO_2 and the other trace gases. High-altitude aerosols, such as those injected into the stratosphere by volcanic eruptions, tend to reduce solar radiation reaching the lower atmosphere, causing a net climatic cooling. An increase in stratospheric sulfuric aerosols from increased carbonyl sulfide (COS) emissions would have a similar effect, but with longer term implications (Wang et al., 1986).

The climatic effect of aerosols injected into the troposphere is more difficult to quantify. Such regionally-dependent aerosol types as soot, sulfate, desert soil, Arctic haze, and maritime aerosols may contribute to regional effects on climate. However, trends in the distributions of such aerosols are poorly known.

Aerosols in the troposphere result from a variety of sources. Natural sources over continental areas include combustion (e.g., forest fires), chemical reactions of natural trace gases, soil and dust particles raised from the Earth's surface, and pollen plus volatile hydrocarbons emitted by plants (Prospero et al., 1983). Anthropogenic sources include SO_2 and elemental carbon emissions from combustion processes. Over the oceans, aerosol sources include sea salt and oceanic emission of DMS (Bates et al., 1987). Conversion of DMS to form aerosols is thought to provide the major production of cloud condensation nuclei (CCN) over the ocean; changes in DMS emissions could provide an important feedback on climate through effects on cloudiness and albedo (Charlson et al., 1987; Penner, 1989), as discussed further in Chapter 4.

3.4 *Projections of Future Concentrations*

3.4.1 *Carbon Dioxide*

Projections of atmospheric CO_2 concentrations are highly dependent on projections of CO_2 emissions. The extrapolation of trends in CO_2 emissions when fossil-fuel usage rates were growing at 4.5%/yr led to expectations the atmospheric CO_2 concentration would reach 600 ppmv within a few decades after the year 2000. The rate of growth of fossil-fuel emissions is now expected to be less than 1%/yr. These lower estimates of fossil-fuel CO_2 emissions push the expected (median) date at which a concentration of 600 ppmv for CO_2 would be reached to late in the 21st century and possibly not until the 22nd century (Edmonds et al., 1986). Note that nonfossil-fuel emissions sources, in particular land-use changes, were not included in the uncertainty analysis, and their consideration would increase the uncertainty in estimates of future equivalent CO_2 concentrations. Projecting future emissions is a critical factor for projecting future climate change.

Figure 3.9 displays three scenarios of future fossil-fuel CO_2 emissions considered by Trabalka et al. (1986). These scenarios were produced by solving an energy economics model under alternative demand and fuel-mix assumptions and by varying model parameter values (Edmonds and Reilly, 1983). Figure 3.10 shows the resulting atmospheric CO_2 concentration simulated by a model of carbon turnover in the atmosphere and oceans similar to that of Oeschger et al. (1975).

Given a particular scenario of future fossil-fuel emissions, the reliability of CO_2 projections depends most on estimates of the uptake of atmospheric CO_2 by the oceans. If past CO_2 releases from terrestrial pools were as large as many believe, then either net oceanic uptake is larger than current models imply or there are other net sinks for atmospheric CO_2 that are not being considered in current models. These inconsistencies in the analysis of past changes must be reconciled to eliminate the uncertainty they introduce about rates of future concentration changes.

The models of carbon turnover in the atmosphere and oceans now used to project CO_2 concentration can be viewed as interpreters of oceanographic data, such as radiocarbon distributions, that record aspects of the dynamics of this system bearing on CO_2 uptake. The confidence to be placed in these models depends on the accuracy of those data and on assumptions about the structure and function of the system underlying these models. Skepticism is appropriate in both regards.

The distribution of natural radiocarbon in the oceans is the most widely used tracer for model calibration. There are, however, few carbon-14 measurements prior to the start of releases from nuclear weapons. Observed carbon-14 activities may be higher than would occur under natural conditions. The representation of oceanic turnover by a globally-averaged model is likely too simplified; although slow vertical mixing described by the diffusion equation is a major aspect of ocean carbon dynamics, advective fluxes are probably responsible for substantial transport of carbon from surface waters.

Mixing and circulation in the oceans depend on climate and will change if rising greenhouse gas levels cause significant climate change. This feedback of climate change on the carbon cycle cannot be analyzed using models based on data collected under relatively constant climate.

3.4.2 *Other Greenhouse Gases*

There have been many attempts to estimate future emissions and concentrations of

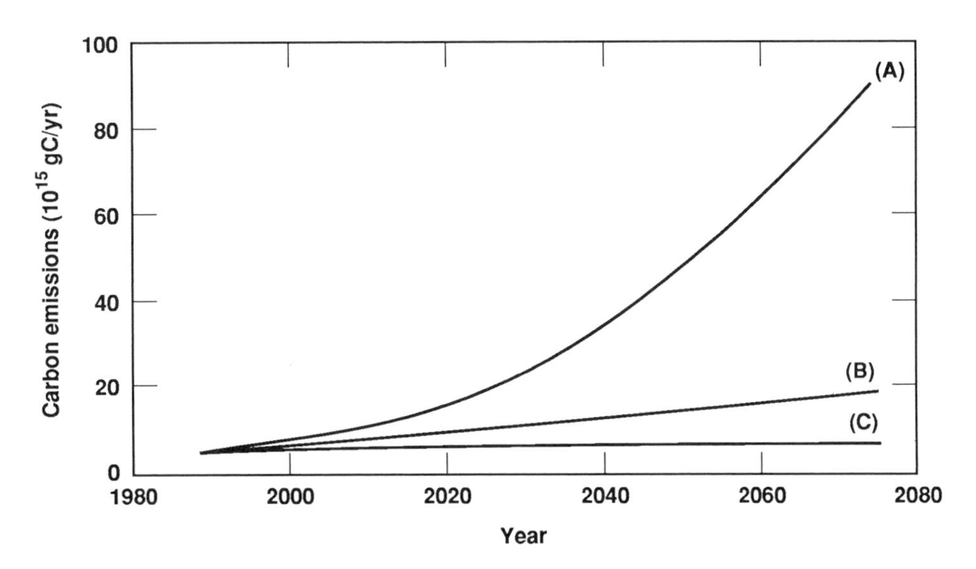

Figure 3.9 Scenarios of future fossil-fuel CO_2 emissions (Trabalka et al., 1986). A, B, and C indicate correspondence with projected atmospheric CO_2 concentrations in Fig. 3.10.

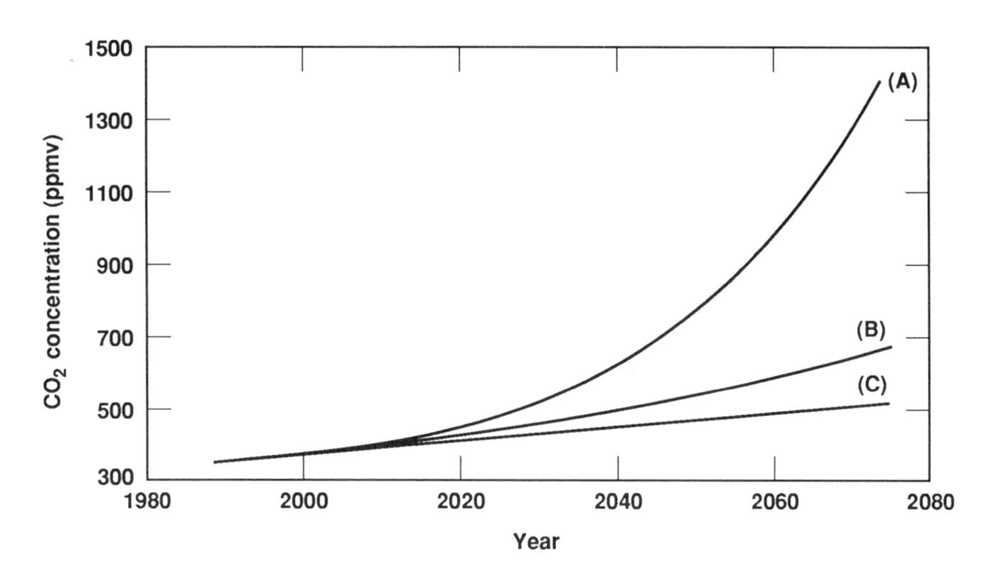

Figure 3.10 Projected atmospheric CO_2 concentrations for the fossil fuel scenarios displayed in Fig. 3.9. These are derived by solving a model of carbon turnover in the atmosphere and oceans that is similar to that described by Oeschger et al. (1975) (Emanuel et al., 1984). A, B, and C indicate correspondence with the fossil fuel scenarios in Fig. 3.9.

Although the development of scenarios for future concentrations of CH_4, CFCs, N_2O, and other gases is useful for theoretical studies evaluating the potential ranges of effects on ozone and climate, such projections are fraught with uncertainties that greatly limit their usefulness.

the non-CO_2 greenhouse gases (e.g., Wuebbles et al., 1984; WMO, 1985, 1989; Mintzer, 1987; Pearman, 1989). Although the development of scenarios for future concentrations of CH_4, CFCs, N_2O, and other gases is useful for theoretical studies evaluating the potential ranges of effects on ozone and climate, such projections are fraught with uncertainties that greatly limit their usefulness.

The factors controlling the increase in atmospheric concentrations of CH_4 and N_2O are still poorly understood; there are still too many uncertainties associated with their budgets to allow us to establish the causes of their increase. For methane, the sources are diverse, making it difficult to establish a global inventory of emissions. In addition, any decrease in the global tropospheric concentration of OH over recent decades would contribute to the net increase in the atmospheric methane concentration.

Although recent measurements suggest that the increase in CH_4 may be slowing (WMO, 1989), there are also several indications that its concentration will continue to increase. Many of the sources of methane are related to human activities; thus, emissions are likely to continue to increase as human population increases. In addition, methane-producing microbial organisms work faster as temperature increases, which would suggest that future climate change could add to emissions of CH_4. Finally, there is the possibility of a massive release from deposits of CH_4 trapped in metastable hydrate formations both beneath terrestrial permafrost and in sediments of the extensive, continental slopes around the Arctic basin and the rest of the world ocean (Revelle, 1983; Cicerone and Oremland, 1988). Although not well quantified, the estimated reserves in these deposits are thought to be many times the present atmospheric levels of CH_4. As the Earth warms from the increasing concentrations of CO_2 and other greenhouse gases, this methane may be released, further adding to the warming.

For N_2O, the recent evidence concerning the overestimate of combustion sources brings into question the actual cause of the increasing concentrations. Generally, scenarios for future assessment studies have assumed that CH_4 and N_2O concentrations will continue to increase at the current rates. It is important to recognize, however, that, because of its relatively short lifetime in the atmosphere, modest reductions in CH_4 emissions could lead to a significant reduction in its concentration.

Projections of future emissions and atmospheric concentrations of CFCs and other halocarbons have been limited by uncertainties about the impact of international regulatory actions. Although the Montreal Protocol currently calls for a 50% decrease in production of these gases by 1998, the high probability of limited participation by developing countries has led to some scenarios assuming essentially constant emissions over coming decades (U.S. EPA, 1989; WMO, 1989). However, current indications are that a near-total ban on the production of CFC-11, CFC-12, and other CFCs may occur as a result of the reevaluation of the Montreal Protocol by the United Nations Environment Programme in late 1989. Although many replacement compounds have been suggested by the chemical industry and have been examined for potential environmental effects by the science community (AFEAS, 1989), the extent of future emissions of these compounds remains uncertain. Most of the compounds being considered have smaller potential effects on ozone and climate than the compounds they are replacing (WMO, 1989; Connell and Wuebbles, 1989).

Although reduced CFC emissions may limit the extent of further decreases in stratospheric ozone, it should be recognized that, because of their long atmospheric lifetimes, existing concentrations of these compounds will continue in the atmosphere for many decades. Significantly reducing the effects of man-made chlorine on stratospheric ozone, particularly over Antarctica, will require that the stratospheric chlorine concentration be reduced below its current level of about 2.7 ppbv (in the upper stratosphere).

Atmospheric model calculations assuming continued increases in the atmospheric concentrations of CH_4, CO, and perhaps NO_x, suggest that the tropospheric ozone concentration could continue to increase, perhaps by as much as 10 to 15% over the next forty years (Thompson et al., 1989a, 1989b; Isaksen and Hov, 1987).

3.5 The Direct Radiative Influence

The contribution of a gas to the greenhouse effect depends on the wavelength at which the gas absorbs radiation, the concentration of the gas, the strength of the absorption per molecule (line strength), and whether other gases absorb at the same wavelengths. Gases absorb and emit radiation at wavelengths corresponding to transitions between discrete energy levels. Absorption at infrared (greenhouse) wavelengths occurs for triatomic or larger molecules where vibrational and rotational energy transitions occur at appropriate wavelengths. Although each transition is associated with a discrete wavelength, the interval over which absorption occurs is "broadened" by the addition or removal of energy as a result of molecular collision (pressure broadening) or the Doppler frequency shift due to the random velocities of molecules (Doppler broadening). If absorption is strong, there may be complete absorption (saturation) around the central wavelength of the spectral line.

As shown in Figure 3.11, at a number of wavelengths the high concentrations of water vapor and carbon dioxide are able to absorb almost completely the radiation emitted at the surface before it is lost to space. Increases in the concentrations of these species would, therefore, lead only to increased absorption in the wings of the absorption lines, with the result that the net trapping of infrared radiation due to these gases would increase logarithmically, not linearly, with concentration. Because the atmospheric temperature changes with altitude, additional concentrations of these gases also change the effective altitude of emission, thereby changing the infrared flux and further enhancing the greenhouse effect of these gases.

Gases absorbing at wavelengths similar to those of CO_2 and H_2O will contribute little to the greenhouse effect, unless they have comparable concentrations. However, Figure 3.11 shows that there is a wavelength region from about 8 to 12 μm where absorption by CO_2 and H_2O is weak; this is referred to as the "window" region. Most of the non-CO_2 gases of interest to climate change, including CH_4, N_2O, O_3, and the CFCs, have absorption lines in the "window" region. Some of these gases, such as CH_4 and N_2O, have absorption lines that overlap other lines. When saturation of line cores occurs, emission from the pressure-broadened Lorentz line wings leads to their absorption increasing approximately as the square root of their concentration. Gases with little overlap, such as CFC-11 and CFC-12, have absorption increasing linearly with concentration.

The net result of these varying radiative characteristics is that comparable increases in concentration of different greenhouse gases have vastly different effects on radiative forcing, as is shown in Table 3.2. The addition of one molecule of methane has about 30 times the effect on climate as the addition of one CO_2 molecule; one CFC-11 molecule is about 22,000 times as effective as one molecule of CO_2. These differences occur because CH_4 and CFC-11 absorb in the "window" region, whereas the CO_2 molecule competes not only with H_2O but also with many other CO_2 molecules (i.e., the 15-μm region of CO_2 absorption is saturated). Although, on a per molecule basis, added CO_2 has the least effect on radiative forcing of the gases considered, it is still the primary gas of concern for climate change because of the larger absolute change in concentration. This is also the case because CO_2 has a longer atmospheric lifetime than the other gases, especially CH_4 making its integral effect over time larger and thereby, in practice, generally reducing the ratios given above.

Ozone plays an important dual role in affecting climate. Whereas the climatic effects of CO_2 and other trace constituents depend primarily on their concentration in the troposphere, the climatic effect of ozone depends on its distribution throughout the troposphere and stratosphere. Ozone and molecular oxygen are the primary absorbers of solar radiation in the atmosphere; absorption of solar radiation by ozone is responsible for the increase in temperature with altitude in the stratosphere. Ozone is also an important absorber of infrared radiation. It is the balance between these radiative processes and the local changes in ozone with altitude (Figure 3.12; also see Lacis et al., 1990) that determine the net effect of ozone on climate. Decreases in ozone above about 30 km tend to increase surface temperature, while decreases in ozone below 30 km tend to decrease surface temperature.

Several research studies using global atmospheric models have attempted to examine the relative effects of other trace gases on radiative forcing and climate as compared to the effects from the increasing carbon dioxide concentrations (Lacis et al., 1981; Hansen et al., 1988; Ramanathan et al., 1985, 1987; Wang et al., 1986). Each of these studies indicates that the combined effects of the other trace gases are comparable to the effects from the projected CO_2 emissions, both for recent observed changes in species abundances and for reasonable assumptions about future changes in their concentrations. For example, model calculations by Lacis et al. (1981) and Hansen et al. (1988) suggest that observed changes in concentrations of CH_4, CFC-11, CFC-12, N_2O, O_3, and other gases during the 1970s and 1980s produced a radiative forcing of 70 to 100% of that expected from the changes in the CO_2 concentration during this period. Results of the Hansen et al. (1988) calculations are shown in Figure 3.13; note that the radiative forcing is only the direct radiative effect on climate and does not account for the additional temperature change expected from the climatic feedbacks associated with clouds, water vapor, sea ice, etc. that may amplify this change by a factor of up to 4. Nor do these results account for the delay of the expected temperature change from ocean-atmosphere interactions. These calculations indicate that by the 1980s, the effect of other greenhouse gases on radiative forcing has become as large as the effect from the increasing CO_2 concentration. Similarly, evaluations of future-scenarios, such as those by Ramanathan et al. (1985, 1987), Wang et al. (1986), and Hansen et al. (1988), suggest that these gases could effectively double, or more than double, the climatic effect from the increasing CO_2 concentration.

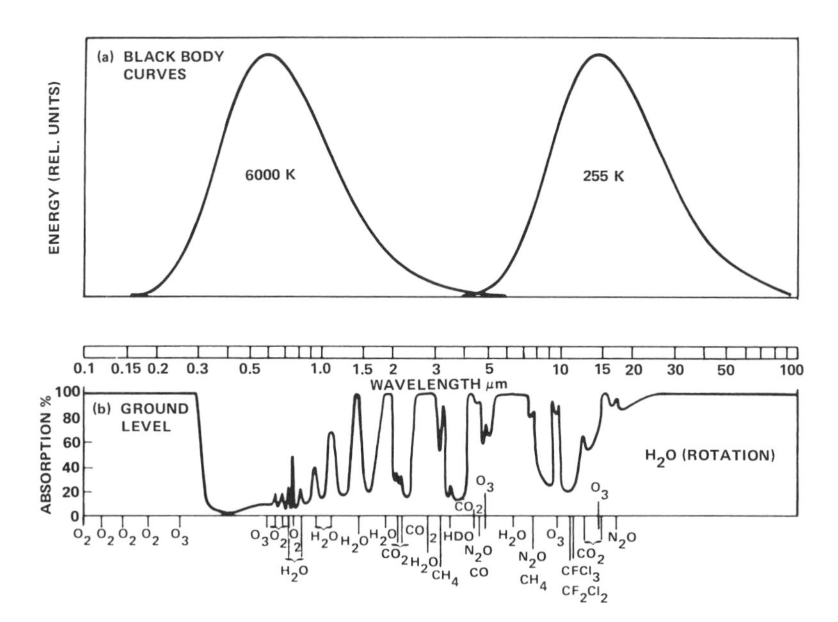

Figure 3.11 (A) Spectral distribution of longwave emission from blackbodies at 6000 K and 255 K, corresponding to the mean emitting temperatures of the Sun and Earth, respectively, and (B) percentage of atmospheric absorption for radiation passing from the top of the atmosphere to the surface. Notice the comparatively weak absorption of the solar spectrum and the region of weak absorption from 8 to 12 μm in the longwave spectrum (updated by MacCracken and Luther, 1985).

Table 3.2 Approximate radiative forcing of various infrared-absorbing gases relative to CO_2 (therefore, $CO_2 = 1$) for each 1 ppbv (or molecule) added to the atmosphere. Based on Ramanathan et al. (1985) and Wuebbles (1989).

Trace gas	Radiative forcing relative to CO_2
CO_2	1
CH_4	30
N_2O	200
CFC-11	22,000
CFC-12	25,000
HCFC-22	7,500
CH_3CCl_3	1,230

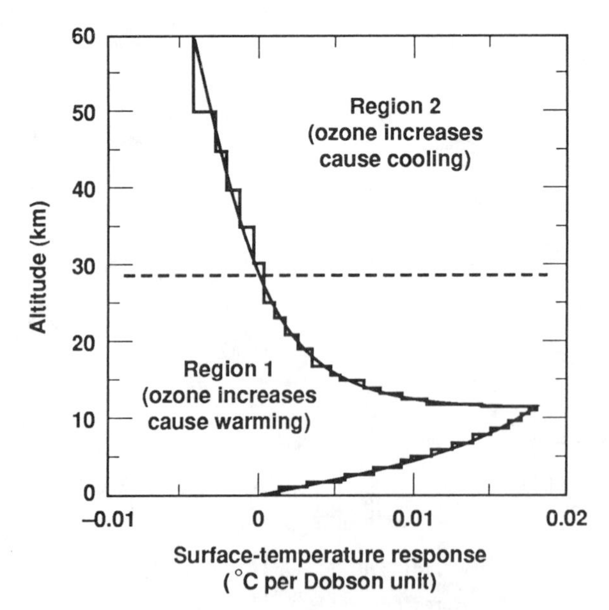

Figure 3.12 Radiative forcing sensitivity of global surface temperature to changes in vertical ozone distribution. The heavy solid line is a least squares fit to one-dimensional model radiative-convective equilibrium results computed for 10 Dobson unit ozone increments added to each atmospheric layer. Ozone increases in Region I (below ~30 km) and ozone decreases in Region II (above ~30 km) warm the surface temperature. No feedback effects are included in the radiative forcing (Lacis et al., 1989).

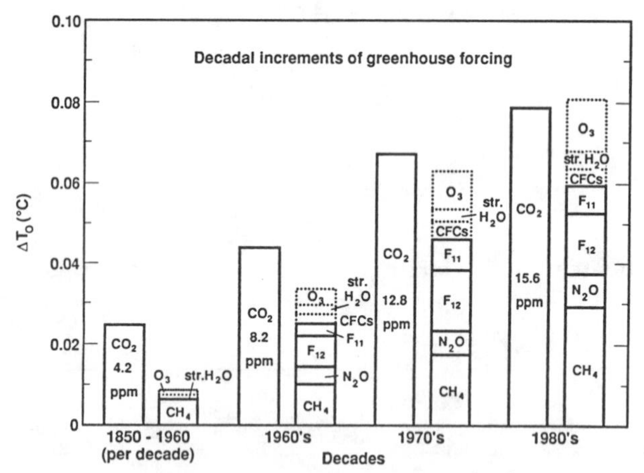

Figure 3.13 Decadal additions to global mean greenhouse forcing of the climate system based on theoretical studies of Hansen et al. (1988). The ΔT_o is the computed commitment to future surface-temperature change once equilibrium is reestablished ($t \to \infty$) for the estimated decadal increases in trace gas abundances, without the amplification by a factor of about 1 to 4 that would be induced by climate feedbacks. Note that, although the CO_2 forcing varies logarithmically with concentration, for the small percentage increments occurring each decade, the effect appears to be nearly linear.

3.6 *Chemical Interactions and Ozone*

As mentioned earlier, the chemistry of the atmosphere is changing as a result of the increased emissions and concentrations of CFCs, CH_4, N_2O, CO, NO_x, and other trace gases. Many of these gases can have and are having an impact on the distribution of ozone in the troposphere and stratosphere. Several trace-gas emission scenarios developed for the recent international ozone assessment study (WMO, 1989) can be used to provide an example of the potential for changes in ozone concentration in the future. The LLNL two-dimensional chemical-radiative-transport model for the troposphere and stratosphere has been used to estimate the resulting changes in ozone concentration.

In all, eight scenarios were developed for the WMO (1989) assessment for emissions from 1980 to 2060. Two of these are described here. In the first scenario, for the Montreal Protocol gases (i.e., CFCs), constant emission is assumed after 1985, using 1986 production as the atmospheric flux. Other gases assume a simple linear change in tropospheric concentration or a compounded rate of increase (CO_2 concentration increases by 0.4%/yr, CH_4 by 15 ppbv/yr, N_2O by 0.25%/yr, CH_3CCl_3 by 4 pptv/yr,[1] and CCl_4 by 1 pptv/yr). The annual rate of HCFC-22 emissions assumed is to increase linearly by 7 million kilograms per year. For CFC emissions, this scenario essentially corresponds to expected global emissions if only the control measures under the existing Montreal Protocol were active.

The second scenario assumes a 95% reduction in production and emission of Montreal Protocol gases, with a 19% reduction per year from 1996 through 2000. HCFC-22 is used as a surrogate for all substitutes for the regulated species.

Figure 3.14 shows the calculated change in total ozone column as a function of latitude and season for 2060 relative to 1980 for the first scenario, with and without temperature feedback (i.e., with and without considering the effects of CO_2 and other greenhouse gases on temperature), using a two-dimensional model (Johnston et al., 1989; WMO, 1989). The largest decreases in ozone occur in the upper stratosphere, where chlorine released from destruction of the CFCs has its largest effect. As mentioned earlier, through its radiative influence, CO_2 tends to ameliorate the effects on ozone of the increasing CFC concentrations. Nonetheless, a significant change in ozone is calculated for this scenario, with largest effects in the polar regions in the late winter and early spring. The effects of heterogeneous chemistry at the winter poles were not included; this would tend to make the decreases in ozone even larger.

Figure 3.15 shows the change in total ozone for the second scenario. These results do not include temperature feedback and should be compared with Figure 3.14a. The calculated decrease in ozone is about half as large as for the first scenario but is still significant. Again, the effects of the Antarctic (and, possibly, Arctic) "ozone hole" would add to the ozone decrease over this period.

3.7 *Research Needs*

The discussion in this chapter, along with that in Chapter 2, clearly indicates that the causes of observed changes in the global concentrations of CO_2, CH_4, and several other globally important atmospheric trace constituents are still not adequately understood. The changes occurring in the sources and sinks of these species must be quantified to determine their ultimate effects on climate

[1] Parts per trillion by volume (pptv) in the atmosphere per year.

a.

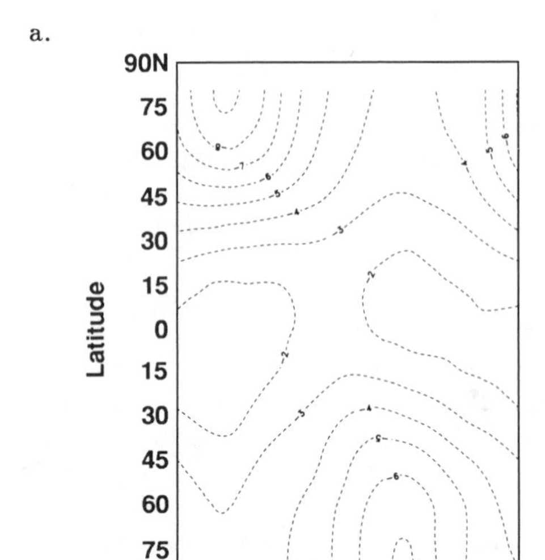

b.

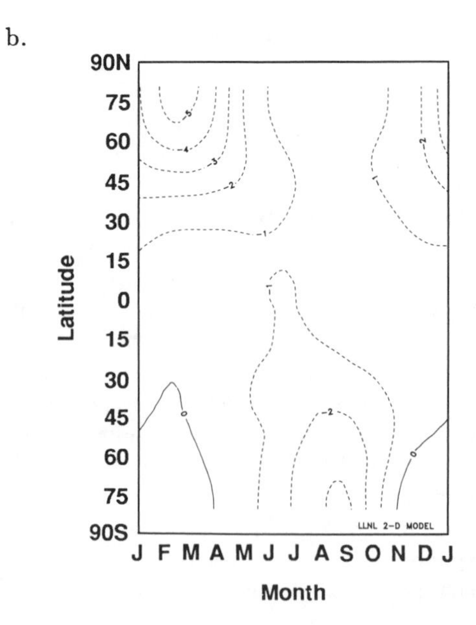

Figure 3.14 Percentage change in total ozone with latitude and season from 1980 to 2060 for the first scenario (constant emissions of Montreal Protocol gases), (a) without and (b) with temperature feedback (WMO, 1989).

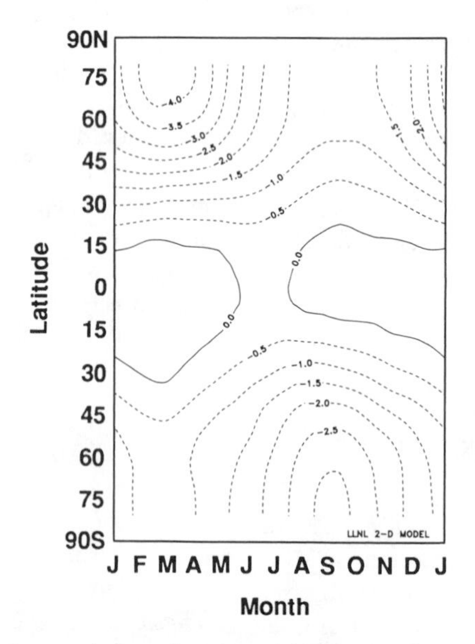

Figure 3.15 Percentage change in total ozone with latitude and season from 1980 to 2060 for the second scenario (95% reduction in Montreal Protocol gases), without temperature feedback (WMO, 1989).

and the global atmosphere. The need for reliable projections of future changes in atmospheric composition also requires the development of appropriate modeling capabilities to simulate the processes determining atmospheric concentrations of relevant trace constituents.

3.7.1 Carbon Dioxide

The CO_2 record retrieved from Antarctic ice resolves much of the uncertainty about the changes in atmospheric CO_2 concentration that occurred during the past 250 years. These data are the best indication of the effects of fossil-fuel use on the atmospheric CO_2 concentration prior to the beginning of direct measurements in 1958; their verification, replication, and interpretation are a high priority, second only to continued monitoring of further atmospheric changes.

Although the atmospheric records from ice and direct measurements document CO_2 changes, these cannot be simply related to fossil-fuel use. We need much better understanding of carbon turnover in the oceans and in terrestrial ecosystems if we are to explain past and project future atmospheric CO_2 changes. As fossil-fuel use continues, oceanic exchanges will likely affect atmospheric CO_2 the most, but changes in carbon stored in vegetation and soil were significant in the past and complicate our accounting of the role of the oceans.

With improvements in measurement precision, it may now be possible, across sufficiently long time periods, to survey changes in the total inorganic carbon content of the oceans, thus establishing the magnitude of the oceanic sink for carbon from the atmosphere. But such surveys cannot contribute as much to our understanding of oceanic carbon dynamics as can be gained by following the redistributions of transient tracers in the sea, particularly radiocarbon and tritium produced by weapons tests. Whether used to calibrate or to test ocean models, these tracer distributions are an irreplaceable register of the dynamics of element cycling in the atmosphere-ocean system.

Ocean mixing and circulation depend on climate, and the ability of the oceans to sequester carbon from the atmosphere may change significantly in response to changes in ocean circulation or marine biosphere activity caused by the projected changes of climate. This important feedback in the carbon cycle and climate systems can only be diagnosed using models that explicitly treat the dependence of ocean mixing and circulation on climate; such models are in early stages of development. A major expansion of work on coupled atmosphere-ocean general circulation models is needed in order to determine the extent to which climate change feeding back on the ocean carbon cycle can exacerbate the greenhouse problem.

Similarly, the processes that cycle carbon in terrestrial ecosystems also depend on environmental conditions. Temperature and moisture are important, as is the atmospheric CO_2 concentration. The increase of the CO_2 concentration may be affecting terrestrial carbon storage already. Direct surveys of global carbon storage in vegetation and soils are impractical—measurement of the small changes that might be caused by global environmental change may be impossible.

A global model is needed that represents vegetation and dead organic matter and can be used to test the consistency of alternative explanations about how environmental effects may compensate for decreases in carbon storage due to land use and other human activities. A major difficulty associated with the formulation of such a model stems from our scant understanding of the factors and processes that dictate carbon allocation to different plant parts, which can determine whether carbon will build up in the soils or simply decay and return to the atmosphere each year.

The transfer to the ocean floor of carbon that marine organisms assimilate is not well quantified. Presumably, marine primary production is limited by light intensity and the availability of major nutrients including nitrogen and phosphorus but not CO_2, which is relatively abundant in seawater. It is important however to measure the accumulation of dead organic matter in order to test nutrient recycling hypotheses that suggest that biological processes are responsible for significant carbon storage in the oceans.

3.7.2 Other Atmospheric Constituents

Budgets of such species as CH_4, CO, N_2O, and NO_x are so poorly understood that we need many additional measurements of the fluxes of these gases from various natural and anthropogenic sources. Biologically-related sources of these and other important constituents must be evaluated for a wide range of areas (e.g., tundra, tropical and temperate forests, coastal wetlands, agricultural biomes) and situations (e.g., biomass burning). Anthropogenic sources also must be measured; for example, the uncertainties in combustion sources of N_2O must be reduced through additional atmospheric and laboratory measurements.

The global distributions of many constituents are also still poorly understood. We must develop improved measurement techniques and global measuring programs. Such data are essential to determine the effects of trace constituent changes on climate and to verify that the models used to study atmospheric composition are accurately representing observed conditions. Overall, there is a dearth of measurements throughout the global atmosphere, particularly in the troposphere. Although the global distributions of longer-lived gases such as CH_4 and N_2O are reasonably known, there exist very few measurements of other important constituents such as OH and NO_x. There are so few stations measuring the vertical distribution of tropospheric ozone that the trend in the global concentration of this important gas is not well established. The role of and concentrations of aerosols also must be better understood.

In the stratosphere, new measurement platforms such as the Upper Atmosphere Research Satellite should contribute to a greatly improved understanding over the next decade. Nevertheless, many uncertainties will remain about the processes controlling ozone unless additional measurement capabilities are developed. For example, the effects of heterogeneous chemical processes on lower stratospheric ozone are still not well understood and will require additional laboratory and atmospheric measurements.

Although progress in the modeling of stratospheric processes through two-dimensional, chemical-radiative-transport models has significantly enhanced the understanding of changes occurring in concentrations of ozone, there remain many areas where modeling capabilities will have to be extended if reliable predictions of future changes in atmospheric composition are to be achieved. Three-dimensional, chemical-radiative-transport models of the global atmosphere have been slow to be developed because of the vast computational resources required. Nonetheless, such models are necessary for understanding the global tropospheric chemical system because of the different source-sink relationships over land and ocean areas. Three-dimensional models would also enhance the understanding of stratospheric processes. Two-dimensional models will remain the primary tool for stratospheric studies and limited tropospheric studies for the next several years, but these models also require additional improvements, particularly in their treatments of radiative and dynamical processes. We must add to these models further treatment of homogeneous and heterogeneous chemical interactions.

Chapter 4: CLIMATIC CONSEQUENCES OF COMPOSITION CHANGES

At the start of the 20th century, the U.S. scientist J.C. Chamberlain (1899) and the Swedish scientist Svante Arrhenius (1896, 1908) recognized that changes in the atmosphere's composition could alter the climate. In the 1930s, the British engineer G.S. Callendar (1938, 1940, 1949) compiled data indicating that both the atmospheric CO_2 concentration and Northern Hemisphere land temperatures were rising, thereby, he claimed, confirming the onset of anthropogenically-induced greenhouse warming. Although broader scientific confirmation of the rise in CO_2 concentration did not emerge until the 1960s (see Ch. 3), confirmation that the global temperature is also rising in accord with projected changes remains largely qualitative even 50 years after Callendar's claims.

> **At the start of the 20th century, the U.S. scientist J.C. Chamberlain (1899) and the Swedish scientist Svante Arrhenius recognized that changes in the atmosphere's composition could alter the climate.**

This chapter reviews our understanding of the climate system and how it operates and then compares the model projections of climate change with changes observed over the past 150 years as a means for evaluating the depth and validity of our understanding.

4.1 Climate and Past Climate Change

4.1.1 Definitions

Weather refers to the instantaneous condition of the atmosphere. *Climate* represents a composite of the weather taken over an appropriate time interval, typically a few decades, and an appropriate spatial interval, ranging from local to global scale domains. This composite includes not only the mean over some particular period (e.g., a month) but also information on variations, range, extremes, and other higher order fluctuations. The National Research Council (NRC, 1975) identified processes (thermal, kinetic, aqueous, and static) and components (atmosphere, hydrosphere, cryosphere, lithosphere, and biosphere) that make up the total system and are responsible for climate and its variations. Thus, the *climatic state* of a region would define "the complete set of atmospheric, hydrospheric, and cryospheric variables over a specified period of time in a specified domain of the earth-atmosphere system" (NRC, 1975). Even though the climate of one January will vary from that of another, if the average over time remains constant, the departure is considered a *variation*; if there is a persistent shift in the average conditions, then we refer to this as a *change* in the climatic state.

In defining climate, the scale as well as the period must be defined (e.g., the annual climate over North America). Shorter time or smaller spatial variations become important descriptors of that climate, and variations over longer periods and larger scales would then look like trends. Areas as large as the entire Earth can be the working spatial scale; time scales of a few decades are generally used to highlight the phenomena associated with the naturally changing climate, man's industrial activities, and the response time of the land-ocean-atmosphere system.

4.1.2 The Climate Record

In order to estimate a change in climate resulting from some external influence such

The Earth's global average temperature appears to have been, with rare exceptions, relatively stable over about the past 5,000 years, generally varying less than about 1°C from its mean over this period.

as alteration of atmospheric composition, it is first necessary to establish a climatic baseline from which the change will occur. This is not a straightforward procedure because the natural phenomena that make up the climate exhibit complex variations on a wide range of time and space scales.

There are many possible measures of past climatic conditions, ranging from recorded temperatures and rainfall in selected locations that extend back a few hundred years to interpretations drawn from surrogate records (e.g., harvest dates) at isolated sites. Examples include profiles of oxygen-18 enhancement in ice cores and sea sediments that reflect the last million years. Theory suggests that carbonates precipitated from a given aqueous solution at different temperatures will contain different levels of oxygen-18, resulting in a correlation between temperature and oxygen-18 levels in the places where both records are available. Paleobotanical studies use pollen cores from undisturbed lake bottoms to determine how the species mix of vegetation in an area varied over time; the method can be used back to about 100,000 years ago. Shorter, but highly detailed, surrogate records are also available from analysis of tree rings, which can provide reasonable spatial coverage back more than a thousand years.

Geological and paleontological evidence clearly demonstrates that the climate has changed, being 5 to 10°C colder much of the last million years, with thick ice sheets covering large areas of North America and Europe, and 5 to 10°C warmer tens to hundreds of millions of years ago when the poles were ice free, swamps and marshes were creating the materials to be made into fossil fuels, and dinosaurs inhabited the planet. There is emerging agreement that a number of factors were primary contributors to past changes of climate, including variations in solar radiance, in the Earth's orbital characteristics, and in atmospheric composition. Understanding is still limited, however, particularly about the relative roles of these factors. Figure 4.1, for example, shows observations of the nearly parallel changes in CO_2 and in the deuterium isotope concentrations (a surrogate for temperature) derived from a deep Antarctic ice core (Barnola et al., 1987; Lorius et al., 1989). Although the agreement is not uniformly good, especially during the sharp temperature drop between about 125,000 and 115,000 years ago, the general correlation does suggest the CO_2 changes may have contributed to the glacial cycling, although they are clearly not the only important factor.[1]

In the absence of changes in these factors, the Earth's global average temperature appears to have been, with rare exceptions, relatively stable over about the past 5,000 years, generally varying less than about

[1] Were the CO_2 variation the only factor contributing to glacial cycling, however, a change of 100 ppmv in CO_2 would be the cause of a 5 to 10°C change in global temperature. Such sensitivity is greater than that currently estimated using climate models. If this observationally-derived estimate were correct, we should probably have seen a few-degree temperature warming over the past 100 years rather than the half degree warming that is roughly consistent with current model estimates.

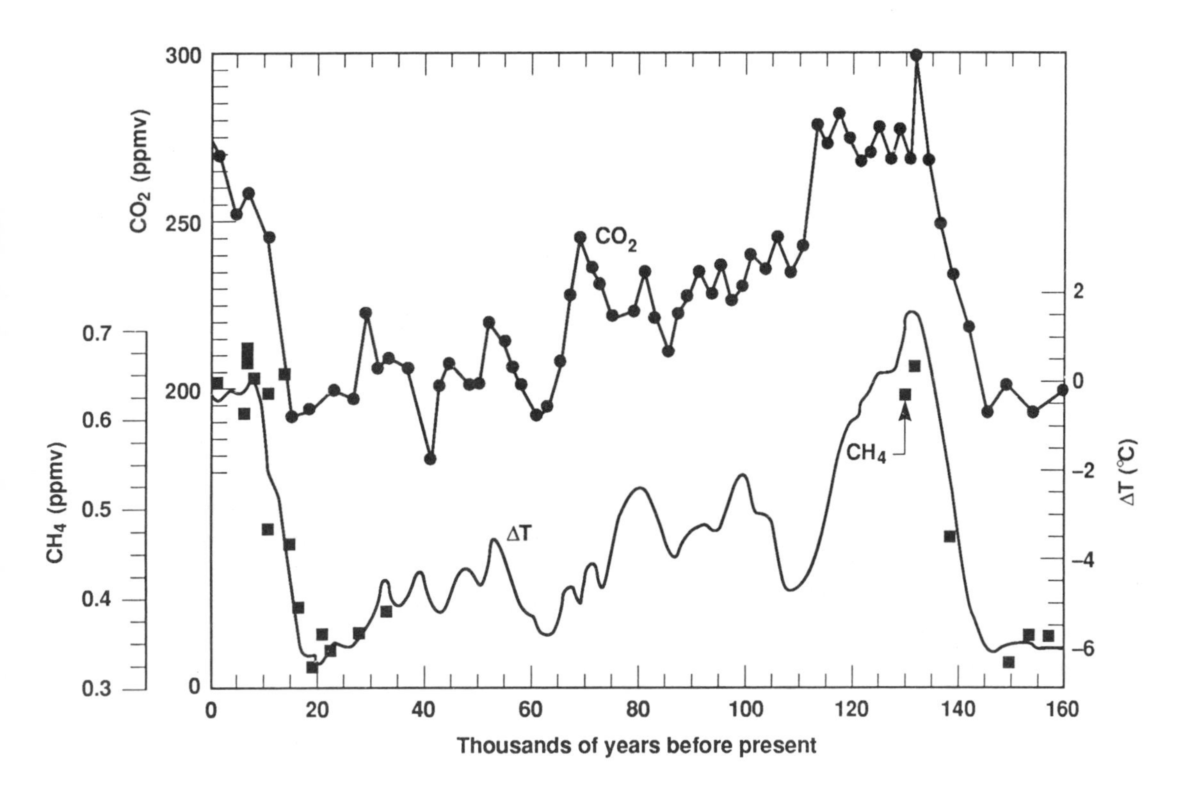

Figure 4.1 Estimates of CO_2 concentration (upper curve, in ppmv) and atmospheric temperature change derived from a deuterium isotopic history (lower curve) for the past 165,000 years from the Vostok ice core. The dark squares indicate concentrations of atmospheric methane (ppmv) from the Vostok or Greenland ice cores (from Lorius et al., 1989).

1°C from its mean over this period. Although the global mean has not varied by much, even relatively small variations can be very important, especially because regional-scale changes can be somewhat larger. One of these events was an extended period generally from about 1600 to 1800 when winter temperatures were particularly cold in Europe, a period that has come to be known as the Little Ice Age. Significant changes in agricultural viability spurred many peoples dependent on the land to migrate, helping populate, for example, the U.S. Information about these early climates comes mainly from use of proxy indicators for the climate such as historical documents on the freezing dates of the Thames River or changes in pollen assemblages preserved in lake deposits. Since the mid-1800s, there has been a generally increasingly complete network of instrumental data. Data sets assembled from the observational record suffer several important limitations, however. For example, measurement methods and techniques have changed, buildings have been constructed near to once rural observation points, and the network does not nearly cover the Earth's surface, especially during the years before World War I and in Southern Hemisphere oceanic and polar regions.

After accounting as carefully as possible for various biases and limitations, analyses indicate that the global average temperature has been rising since the mid-19th century, more uniformly in the Southern Hemisphere than in the Northern (discussed in Section 4.6.2). The trend has also not been uniform over the hemispheres, with some regions experiencing strong warming at the same time that temperatures over the continental U.S., for example, have been relatively constant (Hanson et al., 1989, and Section 4.6.2). Year-to-year fluctuations around the slowly increasing global mean are typically only a few tenths of a degree, and some seem correlated with major volcanic activity and seemingly irregular changes in the tropical ocean circulation (often referred to as El Niño events when the eastern equatorial Pacific Ocean becomes anomalously warm).

Although the very different climates of Mars and Venus make us certain that a change in the atmospheric concentration of greenhouse gases will change the climate, determining by how much and how rapidly the change in the Earth's climate will occur is one of the most formidable problems in physics.

What is most clear from the geological and historical record is that the climate can and has changed on time scales from years to many millions of years. Fluctuations on all time scales probably reflect nonlinear interactions between the system components (atmosphere, ocean, cryosphere, biosphere, lithosphere) that include such processes as mountain building, continental drift, evolution of plant life, and volcanic emission. Some of this variability can be attributed to the complex, nonlinear nature of the climate system, and there may be some chaotic aspects. Whereas weather is not predictable beyond the few days to few weeks that initial conditions dominate atmospheric evolution, the larger fraction of the changes of climate is believed to be potentially determinable. The changing atmospheric composition of greenhouse gases is of comparable magnitude to factors contributing to large changes of climate in the past.

In summary, there is a dynamic background of fluctuations and processes within which man's activities have developed. The proper evaluation of thepotential perturbation of the climate by the increasing concentrations of greenhouse gases must take these natural variations into account.

4.2 *Methods for Projecting Future Climate*

4.2.1 Approaches to Estimating Climate Change

Although the very different climates of Mars and Venus make us certain that a change in the atmospheric concentration of greenhouse gases will change the climate, determining by how much and how rapidly the change in the Earth's climate will occur is one of the most formidable problems in physics. This is so for several reasons.

First, the climate system is a highly interactive, nonlinear system encompassing processes that exhibit variations over a wide range of spatial and temporal scales (see Figure 1.2). For example, cloud processes range from the microphysical ($< 10^{-6}$m), which control cloud reflectivity, to the regional ($> 10^7$m), which determine precipitation patterns; important times of interest in climate studies range from the lifetimes of clouds ($< 10^3$s) to the lifetimes of humans ($> 10^9$s). This broad span of scales greatly limits the ability of scientists to construct physical models in the laboratory capable of investigating overall climatic behavior; study of individual processes, however, can be pursued experimentally, particularly in field experiments using a portion of the ambient atmosphere as the laboratory.

Second, unlike many other such fluid dynamic systems, it is not just the mean state in which we are interested; to understand changes in the statistics of the weather, we must be able to determine the behavior of the turbulent eddies that mix the atmosphere and ocean. It is these eddies that are our weather. While these fluid systems obey the fundamental conservation equations for mass, momentum, and energy, analytical solutions of these equations require such extensive simulations

Although projections indicate that future climate changes will exceed natural variations in the next several decades, we are not yet able to separate the greenhouse signal from the natural fluctuations and are therefore unable to extrapolate with confidence to future patterns of climate change.

that they can only be used to estimate the approximate magnitude of the potential global warming.

As best we can determine, the present rate of change of atmospheric composition change is more rapid than in any past geological period. For this reason, because we can reconstruct only the general patterns of past climates and because of different continental and glacial distributions in the past, past climatic states characterized by strikingly different atmospheric compositions can serve only as rough indicators of the sensitivity of the climate to the composition changes that seem to lie ahead. Although projections indicate that future climate changes will exceed natural variations in the next several decades, we are not yet able to separate the greenhouse signal from the natural fluctuations and are therefore unable to extrapolate with confidence to future patterns of climate change.

With the limitations to approaches based on experiments, theoretical analyses, analogs drawn from the past, and extrapolation, the best available approach is to construct computer-based numerical models of the global climate system that incorporate, in a quantitative way, all of our understanding of the interacting processes. Because simplifications are necessary, even when using the largest available supercomputers, and because we do not have a perfect understanding of the climate, the validity of these models must be evaluated by comparison of model

With the limitations to approaches based on experiments, theoretical analyses, analogs drawn from the past, and extrapolation, the best available approach is to construct computer-based numerical models of the global climate system that incorporate, in a quantitative way, all of our understanding of the interacting processes.

results with past and present climatic behavior before the models are used to project climate change. Although many shortcomings remain, climate models nonetheless represent the best and only means for making comprehensive, self-consistent projections of future climatic conditions over the entire globe.

4.2.2 The Basis and Structure of Climate Models

Climate models represent the global system shown in Fig. 1.1 by dividing the atmosphere, ocean, sea ice, and land surface into grid squares, typically, several hundred kilometers on a side and with vertical extent appropriate to the domain, a few kilometers in the atmosphere, several hundred meters in the ocean, and tens of centimeters to meters for sea ice and land. Fluxes of energy (i.e., temperature), momentum (i.e., winds), and masses of species making up each domain (air, water, and other trace species—gases in the atmosphere, salt in the ocean) are determined by solution of fundamental conservation equations (Washington and Parkinson, 1986).

The complexity of the climate system with its many processes (e.g., radiation, convection, advection, boundary interactions) must all be represented quantitatively. Thus, for example, the representation of solar radiation must treat the changing angle of the Sun through the day and year, the variations of solar energy with wavelength, the spectrally-dependent absorption and scattering by gases, aerosols, and clouds in the atmosphere (including vertical variations of these species and overlap of cloud layers), the different paths of the scattered and direct solar radiation, and cloud absorption and reflection at the surface. Similarly complex representations are required for infrared radiation, vertical convection and mixing, cloud formation and properties, and interactions at interfaces between the air, land, oceans, and ice, including evaporation, precipitation, runoff, snow cover, etc., to name just the most important processes for the atmosphere. The various processes can also interact with each other. Figure 4.2 portrays schematically some of the various interactions that couple the climate system components shown in Fig. 1.1. As an example of what can occur, if the Sun were to start emitting more radiation, the temperature would rise, causing snow and sea ice to melt back, reducing the surface albedo (reflectivity), allowing more solar radiation to be absorbed, and further raising the temperature in a continuing, positive feedback loop. There are many such feedback loops in the system, some positive (amplifying) and others negative (moderating).

Although it is desirable to keep track of all of the major and minor processes, the latter of which are controlling when major processes balance each other, human and computer resources and deficiencies in understanding impose important limits. In some cases only the atmosphere is treated, in others only latitudinal and vertical dimensions are represented, and in other models processes are simplified by, for example, assuming that the Sun is shining continuously, although less intensely, so that fewer radiation calculations need be done as the Sun rises and sets. As different groups have chosen to follow slightly different paths to try to optimize use of resources and to optimally extend scientific understanding, a hierarchy of

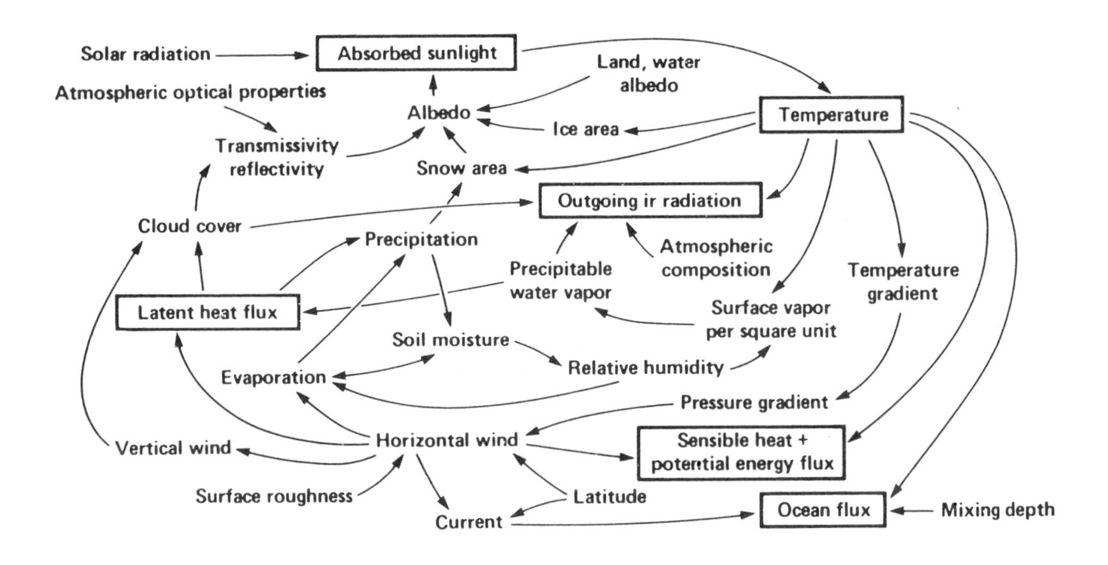

Figure 4.2 Schematic illustration of some of the many interactions that control the global climate (Schneider, 1974). The complexity results from the coupling of many processes that lead to positive and negative feedbacks that can amplify or moderate the climatic consequences of perturbing influences such as the changing atmospheric composition.

models has developed. The most comprehensive of these models are called general circulation models (GCMs), which represent the full global atmosphere, land surface, and, in the most recent versions, oceans and sea ice, although to varying degrees of completeness. Because of the broad space and time scales of the greenhouse issue, GCMs are the leading tools for projecting future conditions, with other types of models used to investigate specific aspects of the issue.

The general characteristics of the four leading GCMs developed in the U.S. are given in Table 4.1. One important measure of the models is their resolution. In their current versions, the horizontal resolution varies from as fine as 4° latitude by 5° longitude to as coarse as 7.8° by 10°. As shown in Figure 4.3, even the 4° × 5° resolution means that the models represent areas the size of Colorado as being homogeneous; that is, having a uniform altitude, temperature, air pressure, precipitation, wind speed, etc. Although finer resolution would be desired, especially in topographically complex areas such as California, each refining of the horizontal and vertical resolution by a factor of 2 (e.g., from 4° × 5° to 2° × 2.5°) increases computer requirements by about a factor of 10; in that about ten hours of supercomputer time are required per simulated year with present resolution, 50-year simulations with significantly higher resolution become very demanding.

Table 4.1 Characteristics of four general circulation models used to project future climatic change.

Host institution	Horizontal resolution (lat. × long.)	Number of layers in vertical	Treat diurnal cycle?	Model mnemonic	Reference
National Center for Atmospheric Research (NCAR)	$4.5^\circ \times 7.5^\circ$	9	No	CCM	Washington and Meehl (1984)
NOAA Geophysical Fluid Dynamics Laboratory (GFDL)	$4.5^\circ \times 7.5^\circ$	9	No	GFDL	Manabe and Wetherald (1987)
NASA Goddard Institute for Space Studies (GISS)	$7.8^\circ \times 10^\circ$	9	Yes	GISS	Hansen et al. (1984)
Oregon State University (OSU)	$4^\circ \times 5^\circ$	2	Yes	OSU	Schlesinger and Zhao (1989)

Note: All models include realistic topography and geography, interactive and multiple-layer cloud cover, variable snow and sea-ice coverage, and interative soil moisture. Treatment of the oceans varies among models and for a given model with the type of simulation.

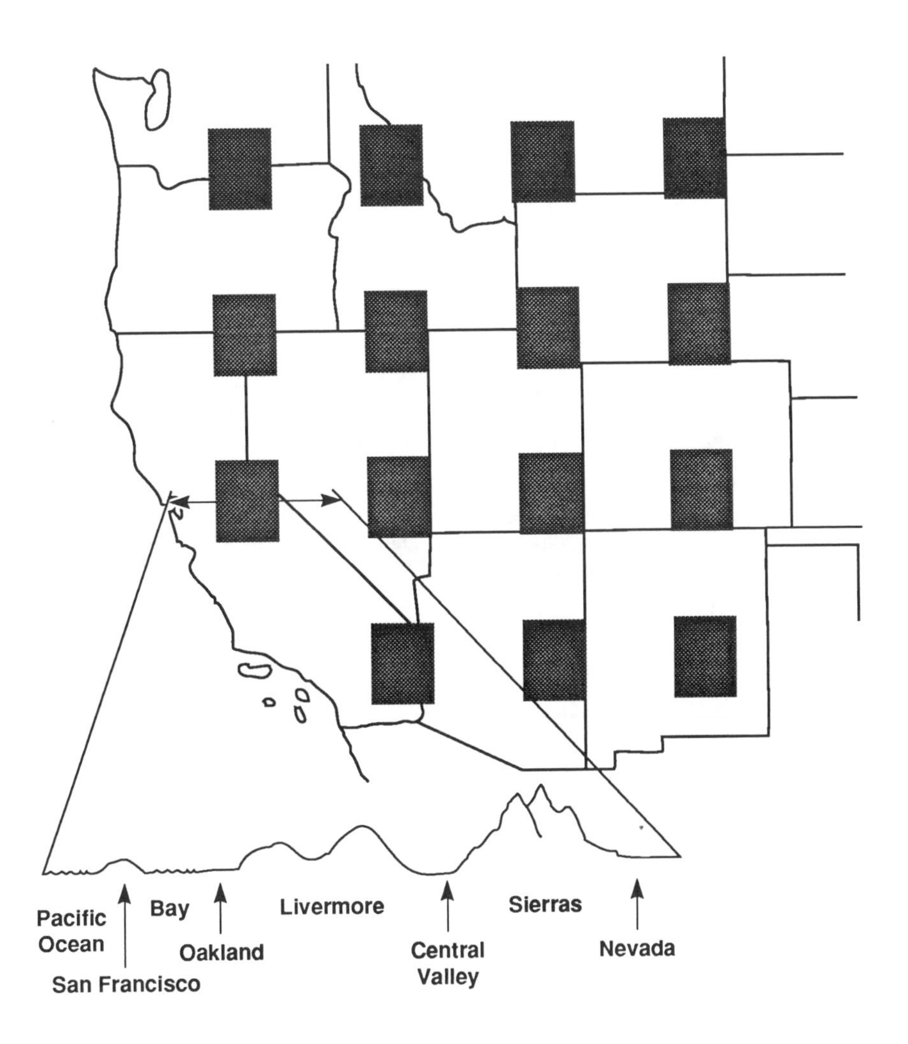

Figure 4.3 Example of gridding used by a general circulation model overlain on the western U.S. The shaded boxes indicate the centers of grid squares when using a 4° latitude by 5° longitude grid, which has been the finest resolution used in recent studies of the potential CO_2 effect by U.S. modeling groups.

Although many shortcomings remain, climate models nonetheless represent the best and only means for making comprehensive, self-consistent projections of future climatic conditions over the entire globe.

Because some processes are too fine-scale to be explicitly represented on the GCM grid scales, however, the net effects of these processes must be represented in terms of parameters and variables on scales that the model does calculate. For example, even though clouds are typically up to a few kilometers in extent, the important radiative and hydrologic processes in which they are involved must be parameterized in terms of the larger-scale atmospheric variables that are calculated in the model. Thus, cloud extent is usually made dependent on relative humidity and vertical stability of the atmosphere. In representing such processes, parameters must often be adjusted within plausible limits so that the observed behavior is reasonably matched. Such "tuning" is essential for many processes because parameters such as surface drag coefficient over a complex topographic and biological region simply cannot be measured.

Given the approximations and simplifications made in implementing the well-established conservation equations, it is essential to test the ability of these models to represent the real climate before using them to project future climatic conditions. This is done in a variety of ways (Table 4.2), ranging from testing individual processes to attempting to represent the large-scale climate changes of the past (e.g., glacial cycling). Because there has never been a "geophysical experiment" similar to that now taking place, such tests can only tell us that the model has not yet seriously failed, not that it is correct.

4.3 Model Representation of the Present Climate

A particularly necessary test of the GCMs is that they represent the present climate accurately. Such a test is not sufficient, of course, because the models have, within plausible limits of parameters, been tuned to reproduce the present climate. Nonetheless, it is instructive to examine how well this can be done and to understand what the limitations of such representations are. Although these models would behave very much like weather forecast models if initialized with current weather conditions, the important tests for climate models are their generation of the multi-year average statistics of the climate (including the mean, the day-to-day and year-to-year variability, the frequency of extreme events, etc.). In the following subsections, intercomparisons both among models and with observed data are used to provide a perspective on the relative behavior of models.

4.3.1 Global and Seasonal Distributions of Climate

For studies of the potential effects of changes in greenhouse gas concentrations on the climate, models that interactively represent the atmosphere, land surface, oceans, biosphere, and cryosphere are needed. During the past ten years, however, the most sophisticated models available have generally only been able to represent the atmosphere, the top 50 to 100 m of the ocean, the top layer of the land surface (with no treatment of plant effects), and snow and sea ice. Only very recently have available models begun to represent ocean circulation and the deeper ocean layers that control the long-term rate of change of climate. In that we live at the interface of the land and atmosphere, the ability

Table 4.2 Examples of means for testing global climate models.

Test representation of individual processes

- Radiation: with laboratory and field experiments.
- Dynamics: with weather forecasts.

Test ability to simulate the present climate

- Latitude-longitude patterns.
- Land-ocean differences.
- Seasonal cycle.

Test ability to simulate climates of the past

- Past 100 years.
- Past 20,000 years.
- Past 100,000,000 years.

Model simulations of present atmospheric behavior are generally able to generate the major circulation features and the seasonal and latitudinal temperature patterns.

of models to represent these conditions is of most interest.

Model simulations of present atmospheric behavior are generally able to generate the major circulation features and the seasonal and latitudinal temperature patterns. Thus, model-derived estimates of global average temperature closely match the observed value; temperatures in the tropics are high and those in polar regions low. The models generate the observed intertropical convergence zone of high precipitation and the very dry areas that create the subtropical deserts. Except to the trained observer, the weather patterns generated by each of these models appear very much like observed conditions, especially for winter periods.

To varying extents, however, this is true because parameters and representations within the models have been adjusted and selected, albeit generally within plausible limits established by observations and field experiments. For example, the prescription for the surface frictional drag coefficient for the complex orographic and vegetational land areas is adjusted, within limits established in idealized field experiments, to achieve reasonable wind speeds. Other choices have been necessary because models may not yet include full and interactive representation of all relevant processes; for example, in models representing only the atmosphere and upper ocean, the depth of the upper ocean layer that provides the heat capacity to buffer the seasonal climate evolution is chosen to assure optimal magnitude and phasing of the annual cycle of temperature. Of most concern, however, are cases in which a representation does not introduce an adjustment to the parameter of a physically-based algorithm but instead calibrates a process to force agreement with observation. An example of this is the derivation of the supposed effects of ocean heat transport on the atmosphere from simulations in which surface ocean temperatures are held fixed—such approaches can very effectively obscure potentially serious shortcomings in model simulations.

Intensive efforts are under way to improve the models and to conduct more thorough and independent characterizations of model behavior. As an example, satellite-derived data on clouds and radiation from the Earth Radiation Budget Experiment (Ramanathan et al., 1989) are being used by more than a dozen modeling groups around the world to evaluate model behavior. This intercomparison study suggests that the models are doing well in representing clear sky conditions but are doing much less well in representing the very important effects of clouds on radiation and climate (Cess et al., 1989).

Because we are interested in the ability of the models to simulate climate change, a number of additional tests of model performance are under way. In response to a hypothetical change in sea surface temperature, results from the models participating in the intercomparison study show a wide range in their estimates of cloud effects, from slightly moderating to strongly amplifying (Cess et al., 1989). Although such results suggest significant uncertainty, there are other indications that the magnitude of model responses is correct. These emerging indications include the ability of some models to simulate a variety of situations ranging from climatic fluctuations, such as the Southern Oscillation (Sperber et al., 1987), to long-term changes of climate, such as glacial and interglacial conditions.

Despite the important progress in developing and improving these models over the past few decades, much remains to be done. Although available models are generally able to represent the broad-scale features of the climate, there does not yet exist a fully coupled ocean-atmosphere-land surface model that can represent present and past global scale climatic behavior with high accuracy and without arbitrary constraints. Development of such a next generation climate model should be an important research priority.

Although available models are generally able to represent the broad-scale features of the climate, there does not yet exist a fully coupled ocean-atmosphere-land surface model that can represent present and past global scale climatic behavior with high accuracy and without arbitrary constraints.

4.3.2 Representation of the Regional Climate and Climatic Variability

The most important interactions of climate and societal activities occur on the regional scale and as a consequence of low probability but high impact events such as hurricanes and droughts. Moving from general concerns about a changing climate to a more specific evaluation of potential impacts thus requires that projections be available on relatively fine space and time scales. Demonstration that models are capable of representing the present climate in such detail is therefore an important test of their potential ability to represent climate change.

The limited horizontal resolution of the models currently available imposes important limitations, even if the models were to be perfectly representative. When gridpoints are the size of the state of Colorado, assuming all locations from Denver to Grand Junction or from the Pacific Ocean across central California to Reno have the same temperature, wind speed, and precipitation, it is not at all self-evident how to compare model results and data. In addition, it is generally agreed that model results are only valid on scales of several grid points, raising typical dimensions to 1000 km or more. With such poor resolution, there is no way for currently available climate models to resolve important features such as the ability of the Sierras to draw snow from the atmosphere or the effects of the Great Lakes on regional climate or to simulate severe, but very localized, storms such as hurricanes. Climate models, however, do have the potential to resolve important subcontinental climatic features such as the monsoon, winter storms, and summer dry spells.

Comparison of the ability of models to represent the longitudinal departures from the latitudinal average climate do indicate that models can represent the seasonal shift in land-ocean temperature contrasts and the development of winter storm systems. Figure 4.4 compares the representation of the average winter temperatures over North America for the four GCMs cited earlier. These results suggest that agreement is better on their representation of large-scale than small-scale features. Figure 4.5 compares modeled and observed distributions of temperature over the western, central, and eastern thirds of the U.S. for winter and summer, clearly indicating that the models do better at predicting the medians, maxima, and minima in winter than in summer.

Very initial studies of the ability of models to simulate interannual variability of temperature also indicate that model representations are better in winter than in summer (Grotch, 1988). This seasonal bias is somewhat troubling because it suggests that the stabilizing processes present under warm season conditions are not fully represented, perhaps indicating that the models are not properly including negative feedback processes that might moderate greenhouse warming. In that it is the ability to simulate summertime variability that will allow consideration of changes in drought frequency, this shortcoming (found so far in one of the two models thus tested) is particularly important.

Comparisons of model results to observations for precipitation are even more problematic than for temperature because of the significant spatial heterogeneity of precipitation and the poor data coverage, particularly over the oceans. Again, as for temperature, the models appear to be able to represent wintertime better than summertime precipitation. Thus, the models do generate the wintertime maxima in precipitation along the west and off the east coasts of the U.S., but the models do not seem to represent the relative maxima in summertime precipitation in the southeastern U.S. Similarly, over the world, the models appear to generate development of major storm systems but are less able to represent convective rainfall during the warm seasons.

A summary of the relative strengths and weaknesses of model performance is given in Table 4.3. A troubling aspect is that models do worst at representing the summertime and high-frequency phenomena that have most impact on agriculture, water resources, and other activities, and best on the wintertime conditions that probably have least impact.

4.4 Sensitivity of Climate to Changes in Atmospheric Composition

Past climates have been both significantly warmer and significantly colder than the present climate. Correlations and associations derived from geological evidence (e.g., see discussion of Figure 4.1) indicate that these changes are largely the consequence of changes

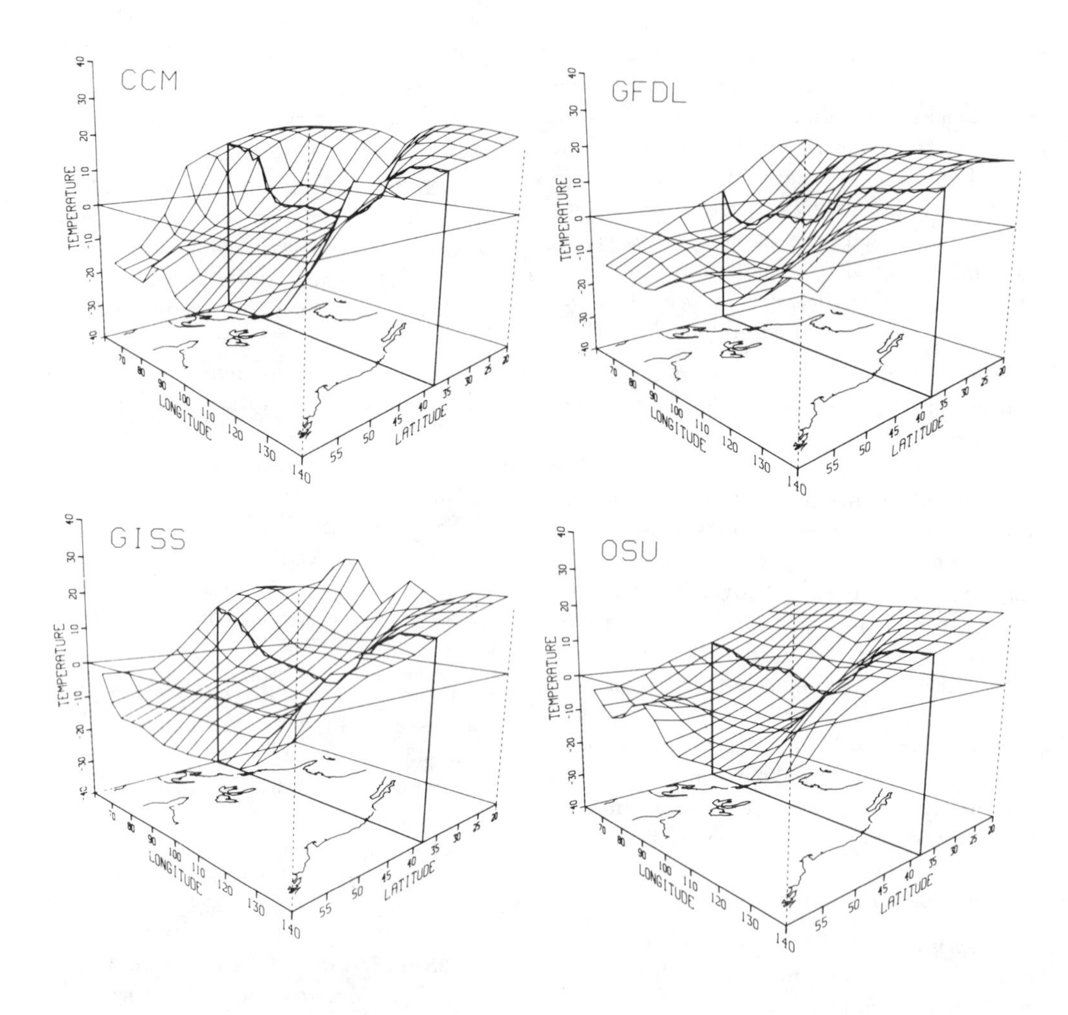

Figure 4.4 Perspective view comparing model estimates of surface temperature (vertical axis) over North America (viewed from Alaska) averaged over December-January-February and showing the differing intensities of land-ocean temperature contrasts for the four models described in Table 4.1 (Grotch, 1988).

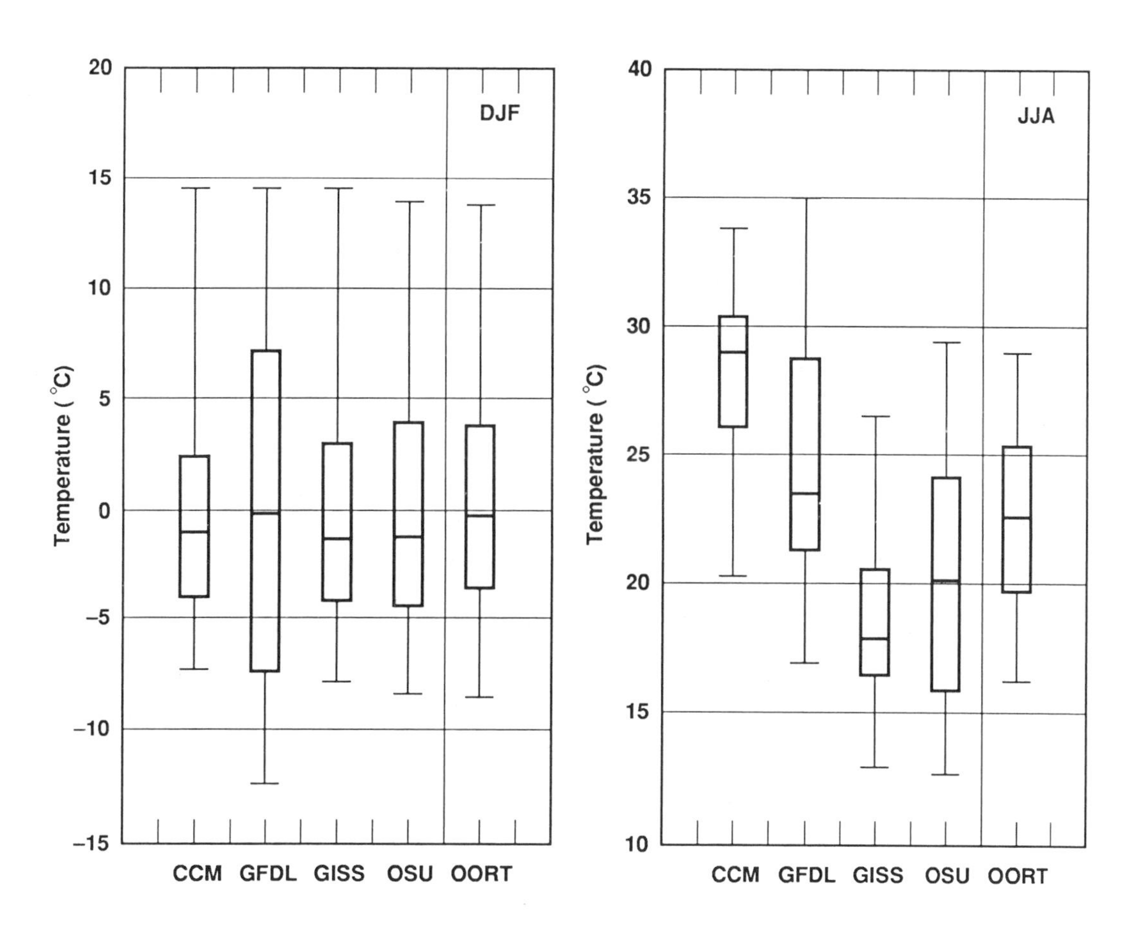

Figure 4.5 Boxplots showing the distributions of surface air temperature estimates (i.e., maximum, 75%, median, 25%, and minimum) for the four general circulation models listed in Table 4.1 and based on observations from Oort (1983) for the gridpoints over the contiguous U.S. Note that the scales for the December-January-February (left panel) and June-July-August panel are scaled differently (Grotch, 1988).

Table 4.3 Strengths and weaknesses of model simulations of the present climate.

Models do relatively well in representing:

- Global average conditions.
- Latitudinal variation of climate.
- Range of seasonal change.
- Wintertime temperatures.

Models do relatively poorly in representing

- Longitudinal and regional variations of climate.
- Summertime temperatures.
- Precipitation and surface hydrology.
- Extreme events.

in factors that determine the climate. Model simulations of past warm and cold periods generally reproduce the reconstructed patterns of past climates when account is taken of independently determined changes in solar irradiance, in the Earth's orbit, in atmospheric composition, and in continental and orographic patterns. Such agreement serves both to increase confidence in the ability of the models to simulate climate change and to confirm empirically that changes in atmospheric composition will indeed lead to substantial changes in climate. Even though past climates have changed and factors such as changes in the Earth's orbit and atmospheric composition appear correlated with these changes, the uniqueness of the present situation necessitates the use of models to provide quantitative projections of how the climate will change in the future. The most straightforward model calculation with which to estimate the magnitude and nature of the potential climate change is to compare a model simulation in which the atmospheric CO_2 concentration has been increased by an arbitrary amount (e.g., to double its preindustrial value) to a model simulation in which the concentration is at its base state (e.g., its preindustrial value). For each of these simulations, it is assumed that a new climatic equilibrium will result, and the simulations are carried out for a time sufficient for equilibrium to be established.

The time for climatic equilibrium to be established depends on the thermal time constants of the system. Therefore, using models representing only the upper ocean layer actively involved in seasonal heat exchange rather than the full ocean shortens the time to be simulated from millenia to decades; this is equivalent to going from years to weeks of computer time, assuming exclusive use of a supercomputer. This saving of computer time and the associated simplification possible in representing the ocean have made it possible for several models to conduct these sensitivity studies, which is important because we cannot totally validate the models and must, at least to some extent, rely on

Model representations are better in winter than in summer. This seasonal bias is somewhat troubling because it suggests that the stabilizing processes present under warm season conditions are not fully represented, perhaps indicating that the models are not properly including negative feedback processes that might moderate greenhouse warming.

agreement among different models to establish confidence.[2] Unfortunately, this simplifying assumption for the ocean also means that these simulations cannot tell us anything about the rate of climate change—only what the commitment to future climate change might eventually be from the arbitrary change in composition. The more involved simulations including representation of the full ocean and the slowly changing concentrations of CO_2 and other trace gases are described in Section 4.5.

4.4.1 Climatic Response to Doubled Carbon Dioxide

Although the atmospheric composition of many species is changing, and although we might prefer to focus on the effects of atmospheric composition changes smaller than a doubling of CO_2, this is the simulation that has been made most frequently. To aid in the analysis, its magnitude is also intended to assure that the climatic response is significantly larger than changes caused by natural climatic variations. Study of a CO_2 doubling should not, however, lead to the inference that this is as high as the CO_2 concentration could go—burning of all the Earth's coal and other fossil fuels could lead to concentrations 5 to 10 times present levels.[3] To account for the potential effects of non-CO_2 trace gases, the similarity of their effects on infrared radiative fluxes allows a scaling using the relationships of relative radiative impact described in Chapter 3.

Although analytical and simple modeling studies done as far back as the beginning of this century suggested that the climatic response would be a few degrees, the first detailed radiative-convective simulation of the average global behavior was done only about two decades ago (Manabe and Wetherald, 1967). By the late 1970s, three-dimensional global models were becoming available, but these models often were restricted by including simplified geography, limiting cloud and ocean interactions, and not treating the annual cycle of the seasons. National Research Council reviews (NRC, 1979, 1982, 1983) concluded from the early modeling studies that a CO_2 doubling would warm the global average surface temperature by 1.5 to 4.5°C. Since that time, the physical comprehensiveness of the models has been further improved, although simplified representations of oceans, the hydrologic cycle (clouds, convection, and soil moisture), chemistry, and in some of the models, the diurnal cycle still remain. The results of these newer models are tending to cluster in the upper half of the NRC range, that is, from about 3 to 4.5°C (Figure 4.6). A recent simulation using a model developed by the United Kingdom Meteorological Office (Wilson and Mitchell, 1987) even suggests a sensitivity of over 5°C. However, as they added additional physics to their cloud algorithm, initially the microphysics of cloud

[2] Even perfect agreement among different models does not assure a correct answer—all models could be leaving out an important, as yet unrecognized, process.

[3] Model studies suggest that temperature changes are logarithmically related to CO_2 concentration; thus, each doubling or halving of the CO_2 concentration would induce about the same temperature response.

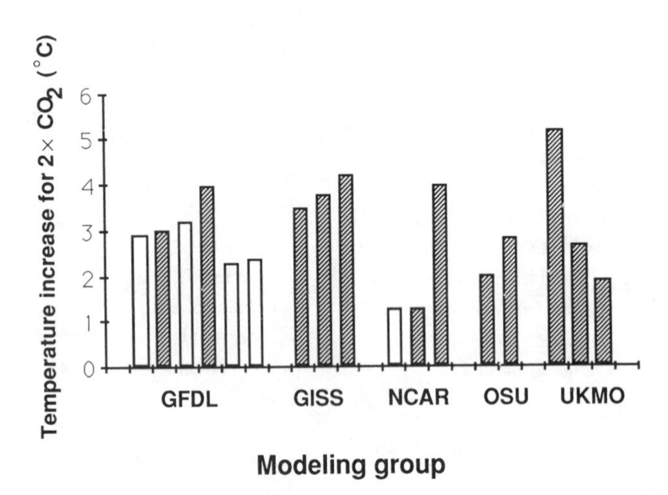

Figure 4.6 Estimated increases in global average temperature resulting from a doubling of the atmospheric CO_2 concentration for various versions of the four climate models described in Table 4.1 and for the climate model developed at the United Kingdom Meteorological Office (UKMO). Open bars indicate model simulations with fixed cloud cover; shaded bars indicate model simulations with variable cloud cover.

hydrology and then the effect of these microphysical changes on cloud optical properties, which would presumably make their algorithm more representative, the sensitivity decreased to less than 2°C (Mitchell et al., 1989). This wide range of results indicates that even ten years of focused research have not narrowed the range of estimates recommended in earlier assessments, in spite of the optimism first expressed in the President's Science Advisory Council report (PSAC, 1965).

Models do worst at representing the summertime and high-frequency phenomena that have most impact on agriculture, water resources, and other activities, and best on the wintertime conditions that probably have least impact.

Our apparent inability over the last decade to estimate global sensitivity to a doubled carbon dioxide concentration arises for several reasons. Our limited understanding of how to estimate the effects of cloud cover and potential changes in these effects is the most important reason. Although the model calculations of clear sky radiative fluxes match satellite observations quite well, their estimates of the amount by which reflection of solar radiation exceeds trapping of infrared radiation span a range of about 100% around the observed value. Although almost all models suggest that warming will lead to an increase in high-level cloud cover that will exceed the reduction in low-level cloud cover, the net effect of cloud cover changes in a simple comparative experiment ranged from a slight moderation of the radiation-induced response to a substantial amplification (Cess et al., 1989), even before the complicating, and potentially moderating, effects of changes in cloud optical properties are also introduced (e.g., Somerville and Remer, 1984; Mitchell et al., 1989).

A comprehensive field and analysis program will be essential to improving understanding of cloud-radiation interactions. A number of other aspects of the simulations also contribute to differences in the model estimates, and thereby to uncertainty in their results. Differences in the treatment of sea ice formation and melting, which creates positive albedo and insulation feedbacks in polar latitudes, cause large differences in estimates of temperature change in high latitudes. Different, and generally highly simplified, treatments of soil water and runoff create substantial differences in the degree of evaporative cooling of particular regions, and thereby large differences in the timing of summer drying and warming. Limitations in the convective prescriptions that remove water vapor from the atmosphere not only create differences in precipitation patterns but also raise questions about the strength of water vapor feedbacks (although Cess et al. found good agreement with model treatment of water vapor in clear sky regions). Differing treatments of the effects of ocean currents, totally neglected in some models, lead to differences in the model representations of the present regional climatic patterns; none of these models contain any representation of the effects of changes in the ocean currents. Until these model limitations and differences are remedied, the range in model estimates of climate sensitivity (one, but not the only, indicator of uncertainty) is likely to remain substantial, although estimates of the global climate sensitivity are not likely to drop below about 1.5°C for a doubling of the CO_2 concentration.

The character of the similarities and differences among the model estimates of climate sensitivity become more evident when the world is divided into four equal areas (Figure 4.7). Temperature changes are larger in high latitudes than in low latitudes, particularly during the cold season in each hemisphere. This amplification is a result of the melting back of sea ice and snow cover, which leads to increased absorption of solar radiation because of the reduced albedo and to increased transfer of heat from the ocean to the atmosphere because of the reduced insulating effect of the sea ice. The different model results also show, however, increasing differences in their estimates of the overall warming.

On even finer scales, the discrepancies tend to be even greater. Grotch (1988) examined the cross-correlations of estimated surface temperature changes over land areas from the different GCMs and found very low spatial coherence among the results, particularly during June-July-August. The largest differences occur for mid-continental regions where the impact of warming on agriculture might be most severe, with the model estimates spanning the range of about 2°C to about 8°C for the U.S. Great Plains. With the exception of these areas and of polar regions where sea ice melts back, the model-estimated range of temperature increase over land areas is generally from about 1.5 to 4.5°C for a CO_2 doubling, with little agreement among models on the distribution of these changes (Schlesinger and Mitchell, 1987). For this reason, studies of potential impacts should consider ranges of possible effects rather than using results from a particular model.

Projections of changes in precipitation and soil moisture are even less certain, particularly for summer periods. Although total global precipitation is projected to increase about 10%, with some tendency to higher percentage increases in the high latitudes, changes are not uniform and some

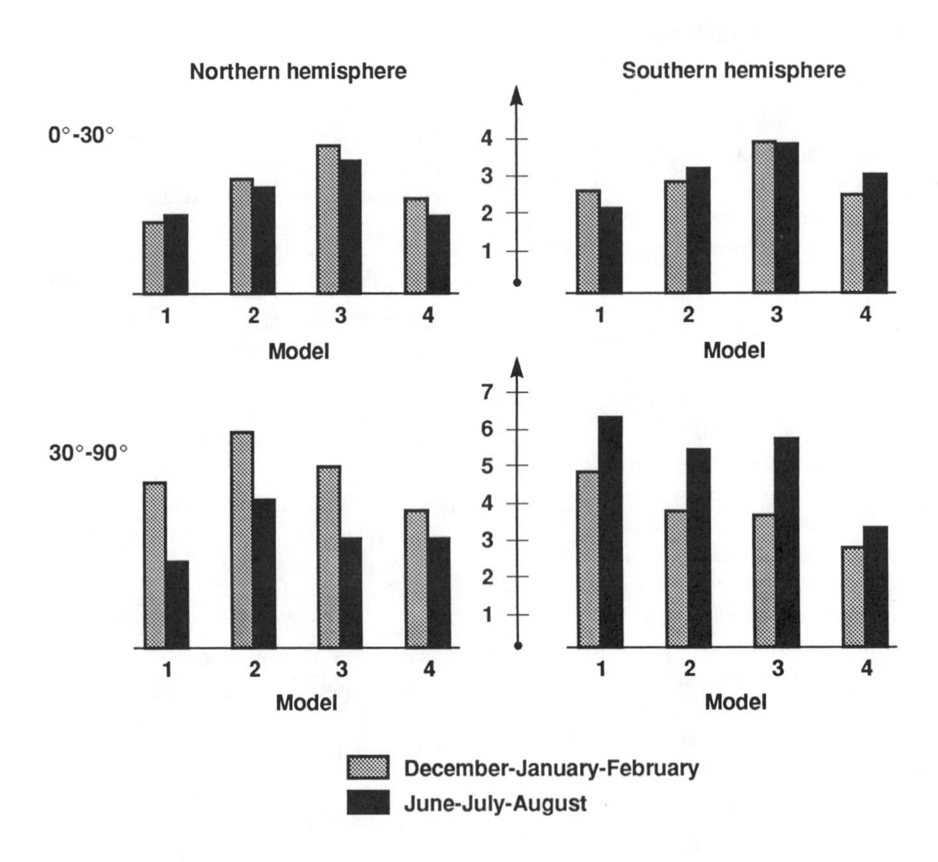

Figure 4.7 Estimates of increase in surface air temperature for a CO_2 doubling from four similar simulations using the four general circulation models listed in Table 4.1 for the low latitudes (0° to 30°) and high latitudes (30° to 90°) for the Northern and Southern Hemispheres and for the December-January-February (left bar) and June-July-August (right bar) seasons.

The increased temperatures will increase evaporation, which may tend to reduce soil moisture in key agricultural regions in summer. The simplified GCM representations of soil moisture, evapotranspiration, runoff, groundwater, and convective precipitation make definitive conclusions about water availability changes premature.

regions will experience decreases. In addition, the increased temperatures will increase evaporation, which may tend to reduce soil moisture in key agricultural regions in summer. The simplified GCM representations of soil moisture, evapotranspiration, runoff, groundwater, and convective precipitation make definitive conclusions about water availability changes premature, however.

There has been only very limited study of model results of changes in climatic variability, maximum and minimum temperatures, drought frequency, and other important climatic elements. Expanded areal extent of warm ocean waters has been used to infer an increase in the frequency and/or intensity of hurricanes (Emanuel, 1987), but changes in wind patterns may also affect both the degree of upwelling of colder waters, either by adding to or moderating this effect, and the availability of nutrients. Available sensitivity studies also provide no information on changing ocean currents, which we know from experience can lead to significant seasonal climatic anomalies.

4.4.2 Perspectives from Past Climate Change

Familiarity with past climate changes and variations provides both perspective for viewing the projected future change and a basis for assessing the validity of the calculated climate sensitivity. As indicated in Section 4.1.2, extended large-scale climatic variations over the past several thousand years have typically been about 1°C or less. Global cooling during the last glacial maximum is estimated to have been about 5°C and global warming during the age of dinosaurs more than 65 million years ago to have been about 5 to 10°C. Thus, even though year-to-year changes of temperature in particular regions for particular seasons can be several degrees or more, the persistence of a climate change as large as a few degrees, as is projected to be the case for a CO_2 doubling, would be significant in a geological perspective, particularly if it occurs rapidly in comparison to the relatively slowly evolving changes of the past (see Section 4.5.2).

Reconstructions of surface temperature and atmospheric composition by Soviet scientists suggest that a doubling of the carbon dioxide concentration from about 300 ppmv to about 600 ppmv might induce a few degree global warming (Budyko et al., 1985). The CO_2 and temperature variations in the Vostok ice core (Figure 4.1) also point to a climate sensitivity of a few degrees Celsius per doubling of the CO_2 concentration (Lorius et al., 1989) when account is also taken of the role of changes in the Earth's orbital elements.

4.5 Projections of Time-Dependent Climate Change

Although the sensitivity of the climate to changes in CO_2 concentration provides an indication of the commitment to future climate change, the actual changes will occur over time as the climate responds to the steady and continuous rise in CO_2 and trace gas concentrations. Although it is often assumed that the consequent climate change will be gradual and continuous, this may not be the case—there are episodes in the past when the climate seems to have jumped from one state to another (see Section 4.6). Developing the ability to project rates of change is critical

The model-estimated range of temperature increase over land areas is generally from about 1.5 to 4.5°C, with little agreement among models on the distribution of these changes. For this reason, studies of potential impacts should consider ranges of possible effects rather than using results from a particular model.

Progress in projecting the rate of climate change lags the ability to project climate sensitivity because there is even less experience in modeling the oceans than the atmosphere, because there are fewer data to verify these models, and because treating these long time constants places large demands on computer resources.

because, for many impacts, it is as much the rate of climate change as its magnitude that is important.

Treating this very complex problem requires representation of the interactions between the upper ocean that buffers seasonal climate change and the deep ocean with its thermal time constant of several centuries. Progress in projecting the rate of climate change lags behind the ability to project climate sensitivity because there is even less experience in modeling the oceans than the atmosphere, because there are fewer data to verify these models, and because treating these long time constants places large demands on computer resources. Only in the last few years have we begun to make significant progress in developing and applying coupled atmosphere-ocean models that properly represent oceanic heat uptake by the upper, intermediate, and deep ocean layers and that can, therefore, be used to estimate the rate of climate change.

4.5.1 Models for Estimating the Rate of Climate Change

Initial attempts to represent the climatic response to the slow rise in CO_2 concentration and the delaying effect on the temperature rise caused by the coupling to the large thermal capacity of the deep ocean were highly simplified. The atmosphere-ocean system was represented as a single, globally-averaged vertical column, and the rate of heat uptake by the ocean was calibrated to match the rate of uptake and downward transport to the deep ocean of various ocean tracers. The transient version of the GISS climate model (Hansen et al., 1988) has extended this approach to treat the entire globe, with spatially dependent diffusion coefficients.

Such pure diffusion (PD) models of the deep ocean focus on treating the perturbation to the deep ocean temperatures, not the absolute value of the temperature change. Because their downward transport of heat is relatively rapid, such models indicate that it takes several centuries for the climate to fully respond to a change in atmospheric composition. In addition, because the vertical mixing rate is dependent on oceanic stability and cooling of the upper ocean surface, this means for calibrating ocean models is quite limiting, especially in estimating the rate of heat uptake (and therefore the rate of climate change).

Another unfortunate shortcoming in applying such PD models in climate models is that at equilibrium the entire ocean depth will have the same temperature, which is contrary to observations indicating a near-freezing deep ocean. This problem with PD models arises because the vertical circulation of the ocean is not being treated.

To overcome these difficulties, upwelling-diffusion (UD) models have been developed. They represent both the downward mixing of heat by diffusive eddy mixing and the upward transport of heat forced by the polar formation of cold deep water. The equilibrium time constant for UD models is several decades, which is significantly shorter than for a PD model.

The coupling of the upper and lower parts of the ocean is, of course, much more complicated than accounted for in these schematic models. Nonetheless, these models do illustrate the need to better represent ocean processes. Attempts have been under way for more than two decades to develop three-dimensional, time-dependent, ocean general circulation models. One difficulty has been that significant horizontal heat transport in the oceans occurs primarily in eddies and currents only tens of kilometers in size, whereas atmospheric storm systems tend to be hundreds of kilometers in extent. A second major difficulty in developing comprehensive ocean circulation models is the lack of continuous, long-term records of temperature, surface wind stress, salinity and fluid motion. Ocean weather stations were a major data source through the late 1970s (Smith and Dobson, 1984) but have largely been abandoned because of the high cost of operation and increased satellite capabilities. The World Ocean Circulation Experiment, scheduled for the 1990s, will provide many measurements essential for improving ocean general circulation models. However, that program will not provide continuous, long-term records, nor does it include measurements of heat, water vapor, and momentum exchanges at the ocean-atmosphere boundary. The latter is especially important for developing a better understanding of coupling between atmospheric and oceanic circulations.

There has, nonetheless, been considerable success in developing models that reproduce many aspects of the observed circulation when observed atmospheric conditions

Even though the atmosphere and ocean models behave reasonably well when decoupled and driven by the observed conditions in the other domain, it has proven necessary in most of the current models to constrain the coupling in often somewhat arbitrary ways to keep the model-simulated climates close to observations.

are imposed at the air-sea interface. An important weakness, however, remains the inability of these models to form cold, deep water, in part because of the necessity to treat the complex, small-scale features involving sea ice formation, brine rejection, etc. For climate calculations, the problems of bottom-water formation and of coupling to the atmosphere must both be addressed before we can generate accurate estimates of the rate of climate change.

Over the past few years, an increasing number of groups have started coupling atmosphere and ocean GCMs for use in climate studies. In contrast to earlier attempts to couple such models at infrequent intervals, the present versions account for the continuous coupling that is necessary to properly account for atmosphere-ocean feedbacks. Ideally, when run for extended periods, these models should evolve into conditions representing the present regimes of the atmosphere and ocean. Unfortunately, even though the atmosphere and ocean models behave reasonably well when decoupled and driven by the observed conditions in the other domain, it has proven necessary in most of the current models to constrain the coupling in often somewhat arbitrary ways to keep the model-simulated climates close to observations. This suggests that further model improvements are necessary in the representation of atmospheric and oceanic physics and coupling.

In exploring how best to constrain such models, the group at GFDL has found that there are apparently two possible circulation states for the ocean and, consequently, different climates for the nearby continental regions (Manabe and Stouffer, 1988). The atmosphere-upper ocean model at GISS also suggests that the potential for natural variations may be larger than we expect, their control climate having shown an unexplained several-year cooling of several tenths of a degree with no change in atmospheric composition (Hansen et al., 1988). Such an apparent intransitivity in the climate implies the potential for a flipping between different equilibrium climates. If such changes are possible—and there are some indications that it may have occurred over a few decades or less as we emerged from the last glacial—it raises the prospect that future climate change may not be gradual and continuous but relatively large and sudden. The potential for such climatic surprises ahead is largely unexplored, although the possibility of such changes is of even more concern since the sudden development of the Antarctic ozone hole.

4.5.2 Time-Dependent Calculations of Climate Change

Simulations are just beginning that attempt to realistically account for both the slow rise in CO_2 and trace gas concentrations and the delaying effects of deep ocean capacity. Hansen et al. (1988) have conducted a simulation with a model having a purely diffusive deep ocean and allowing no change in the heat transported by surface ocean currents. After running their model to simulate about a hundred years, in order to generate a baseline climate assuming 1958 concentrations of CO_2 and other trace gases,[4] they restarted the calculation with concentrations increasing to the observed levels for 1988 and then continuing to increase into the 21st century based on three different emissions scenarios for the future. Their results suggest a warming, compared to the 1950s, of about 0.5°C in the 1990s and, for the continued emissions-growth scenario, up to 1.5°C by about 2020 and 4°C by 2060. As was indicated by the sensitivity studies, the temperature increases tend to be larger at high rather than at low latitudes, but because the natural variability of high-latitude temperature is so high, the change will initially be most evident in mid and low latitudes where variability is generally smaller. Temperature changes were generally symmetric with latitude in these simulations.

This GISS simulation, however, uses a model with fixed ocean currents and only a diffusive representation of the deep ocean. More recent simulations by other groups are beginning to suggest that, just as a UD model gives different results than a PD model, full treatment of the oceans will lead to different geographical patterns of the results. Simulations using the NCAR coupled ocean-atmosphere GCM and including a steady rise in CO_2 concentration (taken arbitrarily to be 1% per year rather than following an emissions scenario), give a quite different geographical pattern of response (Washington and Meehl, 1989). These changes also appear to suggest that global warming may be slower than indicated by the GISS results with a simply diffusive ocean.

These early simulations must still be viewed with considerable caution. Changes in the details of results are likely to occur just

[4] This time was chosen because of the initial availability of reliable data on atmospheric composition. As a result of this choice, however, this simulation and others neglect the warming that would be occurring during this period as a result of earlier increases in the concentrations of the greenhouse gases.

In spite of the uncertainties, however, there are no indications from model and empirical studies that estimates of global-scale warming of as much as a few degrees during the next century will not be realized if emissions continue to increase.

as they did when atmospheric climate models were improved. One clear problem is how to establish initial conditions, there being no assurance that the oceans should be started as if they were in equilibrium, especially given our proximity to the Little Ice Age cooling that ended early in the last century. Another problem is the constrained coupling that now exists in some models; as long as such artificial constraints must be introduced, estimates of the rate of change of global temperature remain uncertain.

In spite of the uncertainties, however, there are no indications from model and empirical studies that estimates of global-scale warming of as much as a few degrees during the next century will not be realized if emissions continue to increase. Table 4.4 summarizes our state of knowledge concerning potential effects over the next century and the degree of confidence that should be associated with them. Although it is difficult to be precise, the likelihood that the global average temperature increase will be less than the range provided by natural variations over the last hundred years appears to be very small, given scientific understanding of the climatic response to identifiable forcings over the geological past. Even a long string of volcanic eruptions would be unlikely to counterbalance the strong warming that is forecast.

4.5.3 Projections of Sea Level Rise

Melting of glacial ice and thermal expansion of oceanic waters have been projected to increase as climatic warming raises the snowline and warms high-latitude and high-altitude regions. Melting of glacial ice would raise sea level by adding to the mass of ocean waters. There is also the possibility, however, that more snow may fall on some new very glacial areas (e.g., the Antarctic), removing water from the ocean and tending to lower (or reduce the increase in) sea level. As ocean waters warm, their density would decrease and volume increase, causing sea level to rise, just as warming a thermometer causes the fluid inside (e.g., mercury) to expand and its level to rise.

Geological evidence confirms that sea level can change by significant amounts. The sea level dropped about 100 m below current levels during the peak of the last glacial advance 18,000 years ago. It was perhaps 5 m above current levels during the peak of the last interglacial about 125,000 years ago when global temperatures are believed to have been about 1 to 2°C warmer than at present. Quite certainly, sea level has been lower during colder periods and higher during warmer periods.

Observations of sea level trends are difficult to make because the rate of change has been relatively small over the last few hundred years (and longer), because coastal areas are not stable platforms, themselves moving up and down for various reasons (e.g., depleting of underground aquifers or oil fields, isostatic adjustments continuing since the melting of the continental ice sheets 18,000 to 12,000 years ago lightened the loading on the continents), and because the ocean tides are comparatively large. After attempting to account for these and other factors, many estimates suggest that sea level rise over the past 100 years has probably been less than about 15 cm, perhaps even less than 10 cm (e.g., PRB, 1985). Some more recent results, however, suggest that the rate may have been as

Table 4.4 Characteristics of future changes in the surface climate.

Reasonably well established

- Global warming by as much as a few degrees through the next century.
- Amplified warming in polar regions as snow cover and sea ice melt back.
- Increased evaporation, especially in summer.

Suggested

- Sea level rise of tens of centimeters.
- Moderated warming at low latitudes.
- Increased summer drying of continental interiors.
- More frequent hot summer days.

Possible

- Increased hurricane frequency and/or intensity.
- Strengthened summer monsoon over southern Asia and Africa.

Uncertain

- Shifting precipitation patterns and storm tracks.

much as twice that amount (Peltier and Tushingham, 1989).

Although some of this change may result from the continuing adjustment of ocean boundaries as a consequence of continental movements and isostatic adjustments, most is believed due to melting of mountain glaciers and thermal expansion of ocean waters. Table 4.5 provides estimates of contributions to sea level rise over the past 100 years from the various possible sources. Melting of mountain glaciers has been reasonably well documented; melting of polar ice sheets is quite uncertain; changes in freshwater aquifers and reservoirs are not listed because they are comparatively much smaller. The estimates for thermal expansion are from the quite simple PD and UD models discussed earlier and are, as a consequence, quite uncertain. Given the broad ranges in both observations and estimates, it is not yet possible to conclude that we have a sufficient quantitative understanding of past changes to assure accurate projections for the future.

Despite these limitations, estimates of future sea level rise have been made. One early method used a PD model to estimate future thermal expansion and then extrapolated to total sea level rise by arbitrarily maintaining the 2:1 to 3:1 ratio for melting of glacial ice to thermal expansion that the authors concluded explained the past rise. This method led to projections of up to about a 3-m sea level rise by 2100, or an average rate of rise of up to about 30 times the rate of the past 100 years (Hoffman et al., 1983). Unfortunately, before the problems with this method were identified, public dissemination of this estimate led to extensive concern in coastal

Table 4.5 Contributions to sea level rise.[a]

Contributing factor	Sea level rise if all ice were totally melted (cm)	Estimated contribution to sea level rise over the past 100 years (cm)	Estimated contribution to sea level rise from 1980 to 2100[b] (cm)
Mountain glaciers	30 to 50	2 to 8	10 to 30
Greenland ice sheet	500	-5 to 10	0 to 30
West Antarctic ice sheet	500	-10 to 5	-20 to 50
East Antarctic ice sheet	6500	~ 0	~ 0
Thermal expansion of ocean water	—	4 to 8	10 to 70
Total change calculated		-9 to 31	0 to 180
Observed sea level increase		10 to 25	

[a] Results are dependent on assumptions about ocean uptake of heat, climate sensitivity, and emission trends. The possible contribution of groundwater depletion is not included.

[b] Adapted from studies by the Polar Research Board (1985), Hoffman et al. (1983), Barth and Titus (1984), Revelle (1983), Robin (1986), Frei et al. (1988), and Meier (1990).

Melting of glacial ice and thermal expansion of oceanic waters are projected to increase as climatic warming raises the snowline and warms high-latitude and high-altitude regions.

areas. Rationalizations that such a rise might occur much later in time and should therefore be studied tended to confuse the issue.

More recent methods are attempting to project the contribution of each component to sea level rise separately. Present estimates are that about half of the mountain ice may melt and that some melting of the Greenland ice sheet may occur. Model estimates of the potential warming over Greenland may, however, be too high because of a lack of model resolution of the Greenland ice cap and the additional meltwater that is created may simply seep into the icesheet and refreeze before reaching the ocean. There is also some evidence that the Greenland ice sheet may be thickening rather than thinning (Zwally et al., 1989). Whether this will continue as warming increases is not certain.

Estimates for the contribution due to thermal expansion are much lower for UD than for PD models, leading to significant uncertainty that must be cleared up with more complete ocean models. Estimates of contributions from changes in the West Antarctic ice sheet are uncertain, first, because we cannot yet accurately estimate changes in the Southern Ocean temperatures and in ocean circulation, which may disturb the shelf ice that stabilizes the ice sheets; second, because a modest warming may actually permit more snow to pile up on the ice sheet, reducing sea level (Meier, 1990); and, third, because over the longer term the ice sheet may start to slide into the ocean (PRB, 1985).

Considering all of these factors (Table 4.5), a plausible central estimate is that

A plausible estimate is that global warming may cause sea level to rise by 50 to 100 cm by the year 2100, with an uncertainty of about 50% ...

global warming may cause sea level to rise by 25 to 75 cm by the year 2100, with an uncertainty of about 50%. The rise would likely not be linear and over the near term may continue to be near past rates. Assuming the warming is not halted, further sea level rise would be expected, with a very uncertain potential for a several meter rise over the next several centuries from melting of polar ice sheets alone.

These changes in global sea level would be in addition to any changes taking place because of other local factors, some of which may be of comparable magnitude (i.e., from extraction of groundwater, isostatic adjustment, etc.). Low-lying areas that are now sinking, such as the Mississippi and Sacramento River deltas, would thus be affected sooner than naturally low-lying areas, such as Bangladesh and the Maldive Islands. Although the tides in many areas vary by this amount, the persistence of the rise, particularly during times when storms push water into the coasts, would endanger even the more elevated coastal areas.

4.6 Consistency of Climatic Trends and Projections

If future increases in the atmospheric concentrations of CO_2 and other greenhouse gases are projected to cause warming and other changes of climate, then a test of those results is to compare projections of the changes of climate from recent increases in greenhouse gas concentrations with changes in the observed climate. Such a test is often referred to as *detecting* that the climate has changed. There has been significant scientific controversy and consequently public confusion over the question of detection (e.g., Kerr,

1989). For this reason, it is important to clarify the tests that are being proposed and the criteria that must be met, because imprecise terminology is in large part the cause of the disagreements.

4.6.1 Tests for Detecting Climate Change

One type of test is to analyze the climatic record, as best it can be constructed, to determine if some measure of the climate, for example, the temperature, is changing from an earlier baseline value by an amount that is statistically significant. Such a test, which requires that the change be greater by a statistically significant amount than whatever natural variations may be taking place and be both in the general direction suggested by model simulations and consistent with our understanding of the causes of past changes in climate, allows an *inference* that the climate is changing due to the changes in greenhouse gas concentrations, which, over this period, have been the largest climatic forcing factors of which we are aware.[5]

Performing this inference test requires a good data set and application of appropriate statistical methods. Data sets are unfortunately limited by several factors, including length of record (most records do not extend back past 1900, when changes should already have begun), areal coverage of the measurements (because local variability is larger for small than for large areas, large areal coverage is desired, but complete global coverage is generally not available), problems and biases with the measurements (e.g., changing measurement times, environments, and methods) and other factors. Not having a preindustrial baseline climate also complicates the statistical problem because of possible trends in the early data and because it is harder to remove the contributions of natural variations when a long baseline is not available. In spite of these problems, the ability to demonstrate that climate change is occurring would be an important first step in proving that increases in greenhouse gas concentrations are capable of changing the climate.

The ability to demonstrate that climate change is occurring would be an important first step in proving that increases in greenhouse gas concentrations are capable of changing the climate.

This inferential test does not, however, confirm that the model projections of future change are correct. To draw that conclusion, a more definitive test is required to quantitatively relate the observed changes to model projections of the changes expected over the past period of record, separating out natural variations from the greenhouse-gas-induced effects. This type of detection test requires that a set of model-predicted changes characteristic of greenhouse-gas-induced climate change (variously referred to as a greenhouse "fingerprint" or "signature") be in quantitative agreement with the observed changes of climate. If this fingerprint can be made distinctive for greenhouse gas effects, and unlike that for volcanic and solar influences on climate, the potential exists to perhaps extract some of their contributions to natural variability, making detection of the greenhouse contribution easier and allowing identification of it earlier.

Performing this more stringent *cause-effect* test, however, requires not only high-quality data sets for several climatic measures but also detailed model calculations of expected changes of climate for the industrial

[5] Over geological time periods, changes in orbital parameters, surface albedo, vulcanism, and other factors have likely caused larger changes in climatic forcing than experienced over the past 100 years—and the changes of climate have also been larger.

period. The problems with the data sets mentioned above become even more severe as the set of measures expands. A more severe limitation, however, has been the complete lack of model calculations realistically covering this period, a result not only of limitations in computer resources and model physics but also of uncertainty about how best to initiate and conduct the calculation given the recognized, but poorly understood, contributions to climate changes since 1800 of volcanic injections, solar variations, and recovery from the Little Ice Age cooling (the cause of which is not understood). Several investigators have attempted such cause-effect tests (e.g., Hansen et al., 1981; Gilliland, 1982; for a review of these efforts, see MacCracken, 1983; also Barnett and Schlesinger, 1987). However, they have had to estimate the expected present pattern of greenhouse-induced climate change by interpolating between equilibrium simulations made to determine model sensitivity, even though it has been demonstrated that such results give different response patterns than do time-dependent calculations (see Section 4.5.2). Given these difficulties, there have been no convincing claims that this type of cause-effect detection test has been successful.

4.6.2 Recent Climatic Trends

Although cause-effect tests cannot yet be definitive, several measures of the climate appear to be changing in a manner generally consistent with expectations.

4.6.2.1 Surface Air Temperature

Until recently, the only way to estimate the rate of climate change and its magnitude and pattern over the industrial period has been to interpolate between the equilibrium climatic states generated in climate sensitivity studies. To do this, the infrared radiative flux change due to the increasing CO_2 and other trace gas concentrations is calculated and compared to the W/m^2 change caused by CO_2 doubling. Dickinson and Cicerone (1986) estimate the flux change due to CO_2, methane, CFCs, and N_2O increases to be about 2.2 W/m^2. The lag effects of the oceans have then been accounted for by use of either PD or UD models. Pure diffusion models suggest that the present change should be a quarter to a third of the equilibrium change whereas UD models suggest that changes should be half to two thirds of the equilibrium change. The increase in trace gas concentrations and the estimated CO_2 climate sensitivity of 1.5 to 4.5°C suggest that there should have been a climate warming since 1850 of about 0.5° to 1.5°C, assuming that the temperature change is linear in radiative forcing and depending on the effects of the ocean in delaying the warming.

The most carefully compiled and reviewed record of changes in surface air temperature is shown in Figure 4.8 (Jones et al., 1986); other records of both surface air temperature and sea surface temperature show similar changes and variations (e.g., Hansen and Lebedeff, 1987). The global temperature appears to have risen by about 0.4 to 0.5°C over the past 100 years, with the rise being steadier in the Southern Hemisphere than in the Northern Hemisphere. This is at the lower limit of the expected change, suggesting that either (1) other factors such as volcanic activity or natural variability are at least temporarily counteracting the warming, (2) climate is not linear in radiative forcing, (3) the data are not representative, (4) the oceans are more effective than estimated in slowing or moderating the warming, (5) the models may be overestimating climate sensitivity by about a factor of 2, or, most likely, a combination of these factors (Schneider, 1989).

Of these factors, those concerned with natural climatic variations (including variations resulting from other identifiable factors)

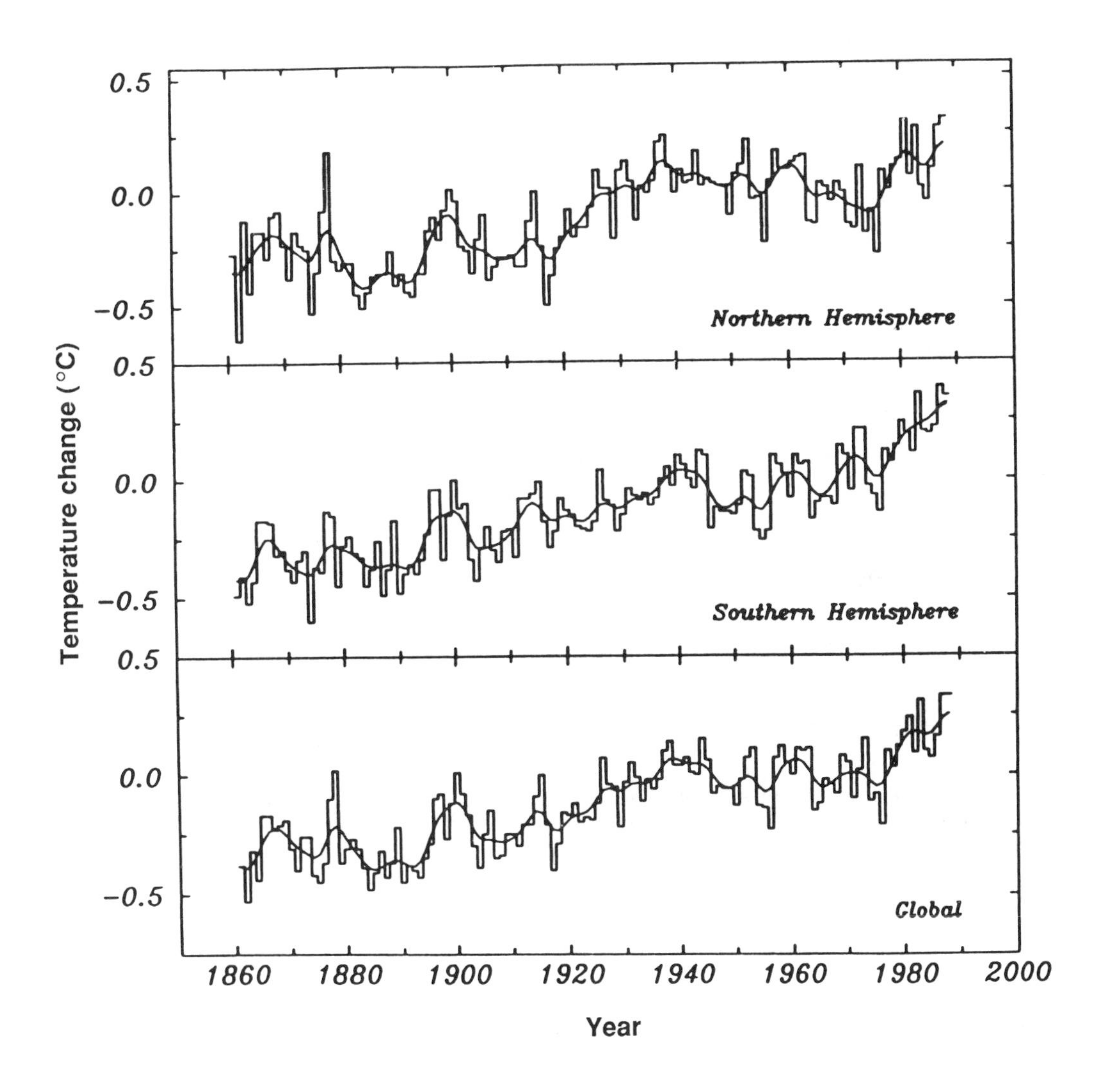

Figure 4.8 Area-weighted estimates of annual temperature departures from a reference normal for the Northern Hemisphere, Southern Hemisphere, and global land and ocean areas for the period since 1860 (Jones et al., 1986).

are particularly uncertain. Over the past thousand years there have been extended periods where there have been warmings and coolings of about the same magnitude as have occurred over the past hundred years. Among the contributing factors to these climatic variations may have been changes in solar radiation, injection of volcanic aerosols, or other factors or natural cycles. There is not sufficient evidence to completely rule out that these factors may be playing some role, but there is no substantial evidence to indicate that these factors alone could be accounting for all of the warming over the past hundred years. This uncertainty about the precise role of natural variability, however, has prevented definitive quantitative identification (i.e., attribution) of the contribution of greenhouse gases to the warming.

There are also other reasons to be cautious in claiming detection. For example, the apparent decrease in the observed Northern Hemisphere temperature from about 1940 to 1970 is particularly troubling, because more of the increase in greenhouse gas concentrations took place during this interval than during the preceding 30-year period when much of the Northern Hemisphere warming apparently took place. A partial explanation for this may be that the earlier warming was a recovery from a period of relatively intense volcanic activity from the 1880s to 1910s. If this is so, then this recovery should not be counted as part of the CO_2-induced warming, and the earlier period prior to 1880 should be used as a baseline. An alternative explanation that has been suggested is that the 1940 to 1970 cooling was a result of changes in ocean circulation, which, if true, makes variability relatively larger and the CO_2 effect harder to distinguish.

A further complication is that the changes have not been globally uniform. Compilations of the temperature record over the U.S. since 1900 (Hanson et al., 1989) show a slight warming in the western third of the country, owing to an increase in minimum temperature (possibly consistent with CO_2 effects, but perhaps an urban effect), but a modest cooling in the eastern third of the U.S. (possibly attributable to cooling by cloudiness changes induced by sulfate aerosols, a change in frequency of warm and cool air mass types, or perhaps just a local fluctuation). That changes over small regions are not consistent with the hemispheric record should not be surprising, but that there may also be inconsistencies in the latitudinal pattern of the changes is of concern. The Northern Hemisphere warming of the 1930s was amplified in high latitudes (consistent with interpolations of model CO_2-sensitivity simulations), but the 1980s warming is greatest in middle latitudes (which could possibly be consistent with time-dependent changes, in that time-dependent simulations show different patterns than do sensitivity simulations). Unfortunately, the full model simulations have not yet been carried out. Also, the lower heat capacity of the Northern Hemisphere land areas compared to Southern Hemisphere ocean areas suggests that the Northern Hemisphere changes should lead those in the Southern Hemisphere; this is, however, not the case. New simulations that represent ocean circulation effects suggest that the Southern Hemisphere warming may lead that in the Northern Hemisphere, but further analysis is clearly necessary. Broader tests of the geographical pattern of temperature changes show similar ambiguities.

4.6.2.2 Ozone and Stratospheric Temperature

The increase in ozone-depleting and greenhouse gases suggests that ozone should be decreasing and stratospheric temperatures cooling because of the increased atmospheric

The global temperature appears to have risen by about 0.4 to 0.5°C over the past 100 years, with the rise being steadier in the Southern Hemisphere than in the Northern Hemisphere. This is at the lower limit of the expected change ... That changes over small regions are not consistent with the hemispheric record should not be surprising, but that there may also be inconsistencies in the latitudinal pattern of the changes is of concern.

emissivity. The stratospheric ozone concentration does appear to be starting to decrease, roughly in accord with model simulations, except in high-latitude regions where heterogeneous aerosol chemistry seems to be creating large depletions that we are only learning how to simulate.

Measurements of stratospheric temperature go back only several decades. There are some indications that the expected cooling is occurring, but the record is short and the changes are quite large and variable, making quantitative confirmation of model simulations difficult.

4.6.2.3 Precipitation

The record of precipitation is limited to land areas and, because of the high spatial variability of precipitation, the record is quite noisy. Generally, the signal being sought is also relatively small compared to natural variability. Nonetheless, there are some indications of diminishing precipitation in lower latitudes and increasing precipitation elsewhere. These changes are in general accord with interpolations from model sensitivity calculations, although not yet convincing.

We can only *infer* that the greenhouse gases have started to change the climate roughly as projected; because we cannot yet quantitatively match model results and observations, we cannot yet *confirm* a definitive and quantitative cause-effect relationship.

4.6.2.4 Sea Level

As indicated in Section 4.5.3, measurements suggest that sea level is slowly rising, with the rate of change within plausible bounds. However, the uncertainties concerning behavior of the polar ice sheets and the rate of ocean heat uptake are so large that the apparent consistency cannot be used to determine whether future changes will be within the estimated range of 50 to 100 cm or higher or lower.

4.6.2.5 Overall Assessment

Taken together, because of limitations in the observational data set and because of uncertainties caused by the lack of accurate model simulations of the expected changes of climate over the past 150 years, we can only *infer* that the greenhouse gases have started to change the climate roughly as projected; because we cannot yet quantitatively match model results and observations, we cannot yet *confirm* a definitive and quantitative cause-effect relationship.

4.7 Research Needs

Our current understanding of climate system behavior is sufficient to warn us that historically significant climate change can be expected through the next century if present emission trends continue. Being able to compare the potential impact of such changes and the many and varied benefits from the activities generating these emissions, if that is to be the measure of whether to take action, requires much more detailed information. Although some of the uncertainties in our understanding will be very difficult, if not impossible, to reduce, many of the aspects limiting quantitative description of the climatic response can be addressed by an aggressive research program focused on improving understanding of atmospheric, oceanic, cryospheric, and biospheric processes; intensified development and verification of climate models; intensive investigation of the natural variability of climate on decadal to millenial scales; and closer comparison of model results to the evolution of climatic conditions over the past 100 years.

The climate system encompasses a wide diversity of processes that interact directly and indirectly in response to the changing atmospheric composition. The initial radiative response to a carbon dioxide increase alone is reasonably well understood, but it is the evolving response in the presence of uncertain changes in clouds, water vapor, ozone, and other factors in a complex and changing convective environment that will determine the actual radiative forcing of the climate system. Studies indicate that it is primarily uncertainties in the cloud-radiation response that determine the magnitude of the global response (Cess et al., 1989). A coupled observation, field, and modeling program is critical to reducing these uncertainties.

Inadequate representation of hydrologic processes, especially the factors controlling increases and decreases in soil moisture, appears to be the primary reason that model results differ on local scales. The water budget at the surface and its interactions with plant growth, evapotranspiration, surface albedo, and other processes must all be better understood if regional-scale impact studies are to be more definitive.

Before we can project the rate of climate change, the heat uptake of the oceans and the responses of sea ice and ocean circulation must be better understood. The World Ocean Circulation Experiment proposes a series of field and modeling programs that should improve understanding of both the horizontal oceanographic processes that are important in determining local climates and variability and the vertical oceanographic processes that control the rate of climate change. A complementary research program concerning the large polar ice sheets is needed to improve estimates of potential sea level change.

An intensified effort is needed to develop the next-generation climate model. Because of limited computer and scientific resources, presently available models do not incorporate adequate representations of many of the features that exert control over the climate. Over the next decade several steps can and must be taken:

1. The model domains must be expanded to include coupled treatments of the atmosphere, ocean, land, ice, and biosphere so that the full range of positive and negative feedbacks can be represented.

2. The chemistry of the atmosphere-ocean-land-biosphere system must be treated in three dimensions and included interactively in climate models, and climate-chemistry feedbacks and the relative effects of different greenhouse gases must be studied.

3. Horizontal resolution must be refined, especially over topographically complex regions, from about several hundred kilometers to less than 50 to 100 km to allow estimation of regional climatic effects.

4. Physically realistic and verified treatments of all relevant climate system processes must be included, particularly cloud-radiation interactions, convection, and hydrology.

5. The causes of differences among climate models and between climate models and observations must be systematically identified and reduced in order to provide more realistic and better tested capabilities.

6. The abilities of the models to simulate past climatic conditions must be thoroughly examined over a wide range of time scales to increase confidence in their ability to estimate the future rate of climate change.

Accelerating the present level of research will require substantial computer resources and ways to improve and optimize their use, including faster hardware, better software, and innovative representations of model physics.

To support all of these model improvements and to confirm their projections of climate change, it is also essential that a comprehensive monitoring, data collection, and analysis program be conducted. Both satellite and surface networks have important roles to play. New instrumentation to detail the atmospheric structure and composition remotely would be immensely valuable.

Although the pace of progress in understanding climate has been steady, the rate of increase in the demand for climate information is much greater than can be matched by the present level of effort (also see CES, 1989). If energy policy must indeed be adjusted to take global-scale climate change into account, the intensity and resources for research on climate change must be increased.

Chapter 5: CONSEQUENCES OF CLIMATE CHANGE FOR THE HUMAN ENVIRONMENT

Climate change will perturb the human environment. This chapter discusses the relationship between development of a national energy strategy and such effects (Section 5.1), compares approaches that have been taken to measure the human consequences of climate change (Section 5.2), and outlines the results of climate-change impact studies that have been performed on individual sectors (Section 5.3) and entire regions (Section 5.4). This chapter also discusses studies of past environmental changes resembling the effects of possible future climate change (Section 5.5) and the major sources of uncertainty (Section 5.6). The chapter concludes with a discussion and summary of effects, knowns, unknowns, and required research (Section 5.7).

5.1 Energy Strategy and the Human Environment

To determine whether any given U.S. energy strategy is proceeding in an environmentally safe manner, it is necessary first to understand its environmental, economic, and social consequences. Different energy strategies imply different potential for both beneficial and adverse effects on the environment, which in turn imply either beneficial or adverse effects on human society. Global climate change has particularly pervasive potential environmental consequences. These potential consequences are global as well as national and local in scope, and have the potential for both beneficial and adverse environmental and socioeconomic effects.

It is not the change of climate itself that may provide a rationale for changes in energy policy. Rather, it is the effect that changes in weather and climate may have on individual resources valued by society—crops, water availability, unmanaged ecosystems, coastal development, and the supply of and demand for energy itself. It is this aspect of climate change that makes it relevant for consideration in developing an energy strategy.

It is not the change of climate itself that may provide a rationale for changes in energy policy. Rather, it is the effect that changes in weather and climate may have on individual resources valued by society—crops, water availability, unmanaged ecosystems, coastal development, and the supply of and demand for energy itself. It is this aspect of climate change that makes it relevant for consideration in developing an energy strategy.

Environmental effects of climate change can be classified as biophysical and socioeconomic (Kates et al., 1985). The effects can be modeled on three different levels. In the simplest of all models, climate affects "exposure units"—places, people, and their activities—leading to a set of ordered consequences. The biophysical side of such a model estimates impacts on physical and non-human biological relationships: precipitation and temperature change, plant growth, sea level—a whole series of so called "first-order" effects. On the basis of the first-order effects, the socioeconomic side of the model then estimates "second-order" operational and economic effects on human enterprises that are dependent on natural resources, such as farms, water districts, forest product firms, etc. Finally, the socioeconomic side of the model translates these second order impacts into wider regional, national, and international societal effects (third-order effects). This simplest level of analysis is probably most useful for identifying mechanisms of climate effects and screening those effects for further detailed study. It does not, however, recognize the feedback reactions of human institutions on the biophysical relationships. Such reactions may largely ameliorate or exacerbate the effects of climate alone (Glantz, 1988).

At a second level of sophistication, climate change can be analyzed as only one of several interacting factors that influence both the human and natural environments (Parry et al., 1988a). For example, the effects of drought on agriculture are more severe under conditions where previous adverse socioeconomic conditions have already strained the reactive capacity of the farm sector. This interactive approach was adopted in an analysis of the effects of drought in the Sahel in Africa in the 1970s (Garcia, 1981). Human reactions are recognized in much of the literature on sea level rise (e.g., Titus, 1988), in the International Institute for Advanced Systems Analysis study of climatic variations and their effect on agriculture (Parry et al., 1988a, 1988b), and in some portions of the EPA report to Congress on the effects of climate change on the U.S. (Smith and Tirpak, 1988). This level of analysis is most useful for identifying social policies that might exacerbate or help alleviate the adverse effects of climate change and that might be implemented to take advantage of beneficial changes.

A still more sophisticated level of analysis is the so-called fully integrated approach, which incorporates not only the feedback effects of second and higher order human responses on the biophysical relationships but also the interactions between sectors at the same order of effects. For example, a change in the level of the Great Lakes (and human reactions to this change) would simultaneously and interactively affect the relationships between irrigated agriculture development, municipal and industrial water supply, recreation activities, transportation, and hydroelectric power. A fully integrated approach was proposed by Callaway et al. (1982), was partially implemented by Parry et al. (1988a, 1988b) for agriculture in some regions, and is now being implemented for a single region in a DOE-sponsored study of the effects of climate change (Cushman et al., 1989). A survey of approaches, methods, and applications of climate impact assessment is provided in Riebsame (1989).

5.2 Approaches to Examining Consequences

Several approaches have been taken in an attempt to anticipate the possible effects of global climate change on the human environment. The development of climate impact assessment has incorporated increasing sophistication in projections of future climate and in applying understanding of the complex relationships between climate and the human environment.

Some of the first assessments of climate impacts were based on arbitrary "scenarios" of climate change that were intended primarily to demonstrate the sensitivity of many resources to such change. For example, Waggoner (1983) examined the effects on U.S. agriculture of a 1°C warmer climate with 10% less precipitation. The next generation of analyses (e.g., Rosenzweig, 1985; Cohen, 1986) took advantage of the temporal detail and spatial explicitness of general circulation models (GCMs)—that is, they used the month-by-month or season-by-season simulations of climate for various points or grid-cells on the surface of the Earth. Most recently, analyses such as that of Wilks (1988) have used weather-generating techniques to produce realistic scenarios of daily weather series from the temporally aggregated GCM output that is made available for climate impact studies. Gates (1985b) suggested the use of spatially nested models or empirical orthogonal functions to relate the spatially coarse (i.e., grid cells of up to 650,000 km^2)

GCM output to local weather, but such techniques have not yet been incorporated in climate impact assessment. In addition, with few minor exceptions these studies have all focused on steady-state changes in climate with today's technologies, rather than on the slowly changing climate that is projected and changing technologies. This assumption introduces considerable uncertainty, perhaps causing overestimation in that natural adaption cannot occur, but possibly underestimation because the difficulties imposed by a changing climate are not considered. There is a general need to examine the consequences of time-dependent warming scenarios under the technological regime that would prevail at the time the warming is expected.

Most of the studies that have been done have focused on particular aspects of individual resource sectors in individual regions, reflecting the interests of different researchers (e.g., Rosenzweig, 1985; Blasing and Solomon, 1984; Cohen, 1986; Gleick, 1987a,b; Pastor and Post, 1988; Miller et al., 1987). These sectoral studies are described in more detail in Section 5.3 of this report. Recently, there have been several attempts to focus on potential climate impact in entire regions, within the U.S. and elsewhere in the world, as described in more detail in Section 5.4.

In addition to studies of potential climate impact based on model-based projections of future climate (typically, the equilibrium climate given a doubling of atmospheric CO_2), another approach has been to examine the actual response of the human environment to past climatic variation, the so-called analog approach. The findings from analog studies of climate impact are described in Section 5.5. Whereas analog studies have the advantage of representing documented responses to actual climate, they also have disadvantages: given that the forcing from CO_2 is different from that causing past climate variation, it is not known whether the past climate patterns will be repeated in the future; because of data limitations, such studies are usually very restricted in their temporal and spatial extent; and because the analog changes are not global (or even national) in extent, many important interactions may be missed.

The state of knowledge regarding the sensitivity of energy systems to climate change is very limited.

5.3 Findings From Sectoral Studies

This section focuses on the results of studies for individual sectors, emphasizing what the studies have taught us about the probable impact of climate change and identifying the major remaining unknowns. Many of the unknowns are common to several of the sectors.

5.3.1 Energy

The state of knowledge regarding the sensitivity of energy systems to climate change is very limited. Jaeger (1983) surveys the subject of climate effects on energy systems in a short chapter, and the U.S. EPA (Smith and Tirpak, 1988) reviews potential effects of climate change on U.S. electric utilities. In general, the field is still in the process of cataloging the nature of CO_2-climate sensitivities and developing crude measures of sensitivity.

Most present short- and medium-term U.S. energy forecast models of the residential and commercial sectors use heating and cooling degree days as the variables to forecast energy demand. Statistically derived measures of this weather sensitivity vary from small (U.S. DOE/EIA, 1989) to relatively large (Klan et al., 1989). However, these models generally do not capture the long-term response of energy consumption patterns to

The production potential of renewable energy supplies, such as solar, wind, ocean thermal energy conversion (OTEC), and biomass, are potentially more sensitive to climate change than are conventional energy supplies.

climate. Long-term residential and commercial winter heating demand likely would decrease with higher average winter temperatures, thereby reducing oil, gas, and electricity consumption. Higher average summer temperatures would mainly increase the demand for electricity as the use of air conditioning increased in the absence of better insulation and window-coatings. Industrial energy demand is largely insensitive to either weather or climate. Statistical analyses have not explored the relationship of transportation energy demand to climate, but warmer winters should increase transport activity and energy demand. Warmer summers should tend to increase the use of mobile air conditioners (as well as the demand for CFCs or CFC substitutes) and energy demand unless there is increased use of window coatings or other conservation measures.

The production of conventional oil, gas, and coal is unlikely to be affected by climate, although less severe winter conditions could lower Arctic region supply costs. A possible exception is that permafrost decay could create problems with Arctic infrastructure such as pipelines (Maxwell and Barrie, 1989). The availability and operation of hydroelectric power could be indirectly affected by climate through precipitation and evaporation patterns. Hydrology and average climate could similarly affect the availability and design of power plant cooling systems.

The production potential of renewable energy supplies, such as solar, wind, ocean thermal energy conversion (OTEC), and biomass, are potentially more sensitive to climate change than are conventional energy supplies. Temperature, cloud cover, wind vectors, and their associated variances affect the production of solar, wind, and OTEC energy. In addition, biomass farms would be affected by hydrology and the ambient CO_2 concentration. It is unclear whether increased CO_2 and changed climate would increase or decrease total biomass energy productivity. Biomass waste as an energy source could be affected by productivity of forests. Average temperature and pressure conditions can also affect the availability of recoverable methane release from landfills.

5.3.2 Agriculture

Current projections of world population and income suggest that, in the 50 to 75 years expected for warming associated with a CO_2 doubling to occur (Bolin et al., 1986), global demand for food and fiber will grow at an annual percentage rate about half that of the last 30 to 40 years (World Bank, 1984). Most current studies of the climate changes resulting from a CO_2 doubling imply that these changes will be gradual in the coming 50 to 75 years. If so, the negative impact would likely not overwhelm the prospective gains in productivity from even modest rates of technical advance (Easterling et al., 1989). This would be especially true if an increased atmospheric CO_2 concentration proves to be as effective a fertilizer in the field as it has proven in laboratory experiments (Strain and Cure, 1985). Thus, it appears likely that the future rate of growth in the worldwide demand for food and fiber can be met (Crosson and Rosenberg, 1989), even with projected changes in climate. However, if climate change is more severe and/or rapid, these projections could turn out to be too optimistic.

Climate changes are likely to alter regional agricultural capacity. The nature of these changes is uncertain because general

It appears likely that the future rate of growth in the worldwide demand for food and fiber can be met, even with projected changes in climate. However, if climate change is more severe and/or rapid, these projections could turn out to be too optimistic.

circulation models (GCMs) disagree about the effects of a CO_2 doubling on regional climates (e.g., Schlesinger and Mitchell, 1985; Grotch, 1988). Shifts in agricultural comparative advantage among regions could present some affected countries with difficult adjustment problems while others such as Canada might benefit (Smit, 1989) from comparative advantage. For example, some of the GCMs indicate both hotter and drier conditions in the mid-latitudes. Studies based on that scenario (e.g., Blasing and Solomon, 1982) conclude that the American cornbelt would shift far to the north and east.

Countries losing agricultural comparative advantage because of climate change can be expected to: (1) protect their agriculture by restricting imports from newly advantaged producers (despite the economic cost); (2) support research to develop appropriate technologies including better adapted crop varieties and animals; or (3) import more of the needed food and fiber while diverting investment from agriculture to other more productive sectors (Easterling et al., 1989). The first option may be invoked for reasons of national food security. The second option requires that national and international agricultural research institutions continue to develop and transfer new varieties and technologies to the farm sector. The third option requires confidence in the stability of the international trading system.

Adjusting agriculture to slow and moderate climate change will, obviously, be simpler than if change is abrupt or if it is large and rapid. Warming beyond that predicted to result from a CO_2 doubling or occurring in a much shorter time frame than the second third of the next century would pose more difficult problems for world agriculture than discussed above. If this occurred, global agricultural capacity would not keep pace with rising demand, with the result that economic and environmental costs would be sharply higher.

Because climate models do not agree on the direction of change in annual precipitation for many regions, the impact on regional water supplies is highly uncertain.

5.3.3 Water Resources

Global warming will accelerate the hydrologic cycle. The resulting increase in average global precipitation and evaporation from a CO_2 doubling is estimated at between 7 and 15% (e.g., Bolin et al., 1986). However, because climate models do not agree even on the direction of change in annual precipitation for many regions, the potential impact on regional water supplies is highly uncertain (Frederick and Gleick, 1989). In areas such as northern California where precipitation is currently dominated by winter snowfall and runoff is dominated by spring snow melt, warmer temperatures could produce dramatic changes in seasonal runoff patterns. Even in the absence of any change in total annual precipitation, the higher temperatures would increase runoff in winter and decrease it in summer because more of the precipitation would come in the form of rain, and the snow would melt earlier in the spring (Gleick, 1987b).

The impact of climate change on water supply is uncertain. GCMs indicate possible changes in average annual precipitation for any given region on the order of plus or minus

20% of the present rainfall once the equilibrium change in climate for a CO_2 doubling is established. Because runoff is essentially the difference between precipitation and evapotranspiration, impact on runoff could be even greater. Where runoff decreased, water quality in streams and rivers would decline unless pollutant loads also decrease. The impact on water demand is also very uncertain. Water use in urban and suburban areas would probably increase with increasing temperature. In agriculture, irrigators would tend to use more water to compensate for higher transpiration rates. Increased transpiration would be compensated somewhat by higher CO_2 levels that increase the plant's resistance to vapor transfer into the air (Rosenberg, 1981). Transpiration rates might also be higher or lower than predicted by increases in temperature as the result of (as yet unpredictable) changes in cloudiness, humidity, and windiness (Martin et al., 1989; Rosenberg et al., 1989).

The relative value of water for alternative uses would likely change. Drinking and domestic uses would remain top priorities, but changes in seasonal and annual supplies might alter the relative benefits of allocating water and reservoir capacity to flood control, power generation, fish habitat, or consumptive uses such as irrigation. Hydroelectric power might become more attractive as a means of both abating the greenhouse effect and adapting to increased power demands that might accompany it (Frederick and Gleick, 1989). However, this would depend upon the availability of suitable water flow at existing sites and/or appropriate new hydroelectric sites, which are becoming increasingly scarce in many parts of the world.

Adaptations to climate change could involve construction of new dams and reservoirs, interbasin transfers of water, development of "unconventional" sources of water such as desalinization, recycling of industrial, municipal, and agricultural waste water, and weather modification. Lacking sufficient guidance on the specifics of future climatic conditions, water managers and planners are unlikely to invest in any of these measures unless factors other than climate change already justify them. However, water managers may already be willing to invest in techniques that improve the operation of existing infrastructure and in research and technological innovations to accomplish this end. The prospect of future climate change might hasten such investments (Frederick and Kneese, 1989).

The projected global warming could accelerate over the next few decades to rates that may outrun the natural rates of forest migration, which occur on millennial time scales.

5.3.4 Forests and Biodiversity

The projected global warming could accelerate over the next few decades to rates that may outrun the natural rates of forest migration, which occur on millennial time scales (Batie and Shugart, 1989). Existing forests would become increasingly stressed and more susceptible to infestation, disease, and, eventually, fire. Gradually, existing forests would be replaced by other forms of vegetation or by forests with a different species mix (Sedjo and Solomon, 1989). Whether forests will experience more rapid growth and/or increased efficiency of water utilization, as has been observed in some nonwoody plants, is unknown. Because rising temperature will increase the respiratory demand of woody tissue, net growth could decrease. Evidence for a CO_2 fertilization effect in forests is lacking.

Tree growth is generally limited by lack of summer warmth in the high latitudes and by heat and lack of water in the mid-latitudes. The effects on forests would probably be modest in the tropics, where temperature changes are expected to be least. With global warming, the boreal forests may migrate northward into the currently unforested tundra, provided there is adequate precipitation. Simulations (Solomon, 1986; Pastor and Post, 1988) indicate that the largest transitions will likely occur at the current border of the boreal and cool temperate zones. Some mid-latitude forests could disappear, especially if the moisture-saving effects of high CO_2 and increases in plant water use efficiency do not materialize. Warming in mountainous terrain would cause species to move to higher elevations (Sedjo and Solomon, 1989). Because of the lack of analysis of feedback effects (e.g., fires, pests), any real predictions of the regional fate of forests is premature.

Rapid change in climate threatens a reduction in biodiversity (Batie and Shugart, 1989). Some existing species of plants and animals would be unable to adapt, being insufficiently mobile to migrate at the rate required for survival (Davis, 1989). Although the economic value of biodiversity is difficult to quantify, it is undoubtedly substantial.

Adaptation of the forest sector to changing climate will not be simple but would likely include earlier harvests of unsuited species and salvage operations in older stands, seeding and thinning (which are costly) in younger stands, and active planting of trees adapted to hotter and drier (or wetter—it is not clear which) climates in harvested stands (Sedjo and Solomon, 1989). Introduction of new varieties or introduction into new regions is a much slower process in forestry than in agriculture. At least in the first decades, adaptation might involve changes in the species mix that could require costly adjustments in the logging and processing industry. Also, the long growing periods for trees adds the economic risk of inappropriate species choice for changing climate conditions, inhibiting investment in trees and mills to process them. The geography of production forestry would change, with some regions becoming increasingly important sources of forest products while others declined. Intense management of forests would be limited to those areas where high-yield plantation forestry could continue to be practiced profitably.

Rapid change in climate threatens a reduction in biodiversity. Some existing species of plants and animals would be unable to adapt, being insufficiently mobile to migrate at the rate required for survival.

5.3.5 Air Quality

Global warming is expected to affect regional and global air quality through a number of primary and secondary mechanisms (see Penner et al., 1989, for a full discussion). In general, it is expected that these influences will make improving air quality more difficult.

Primary interactions, such as direct surface warming and the resultant increased soil and biogenic emissions, were discussed above and are more important from a climate-feedback standpoint than from the standpoint of what is normally considered to be ambient air quality. However, in some areas, such as the eastern U.S. where vegetative sources are abundant, these effects may also be important for air quality. Secondary interactions, which result from meteorological phenomena indirectly related to temperature change, can be expected to have pronounced effects on regional and global air quality. Examples of such interactions include:

- Alteration of the wind-field patterns that carry pollutants from their sources to their ultimate receptors, with resultant

Global warming is expected to affect regional and global air quality through a number of primary and secondary mechanisms. In general, it is expected that these influences will make improving air quality more difficult.

changes in ambient pollutant concentrations.

- Modification of stagnation periods and associated pollutant mixing parameters, with associated changes in pollutant levels.
- Changes of the hydrologic cycle and associated storm climatology, with corresponding changes in the wet removal of pollutants.
- Climatological modifications in levels of the solar actinic flux, which affects photochemical conversion rates (and thus the fate) of many key pollutants.
- Changes in temperature or the frequency of high-temperature occurrences which affect photochemical conversion rates.
- Associated modification of climatic regimes, the associated vegetation, and thus the dry deposition rates of key pollutant species.

Although there is consensus that these features are apt to affect ambient air quality under the scenario conditions suggested by current GCM outputs, few direct or substantive predictions to this effect have appeared in the literature. A number of stochastic models of wet-removal have been published (e.g., Rodhe and Grandell, 1972) that demonstrate the strong general dependency of pollution residence times and concentrations on rainfall statistics. Furthermore, a variety of low-dimensional chemical models (e.g., Bruhl and Crutzen, 1988; Penner et al., 1989) have provided some indication to this effect under clear-air conditions. In general, however, these statistical and low-dimensional deterministic models can be expected to give only a qualitative indication of the consequences of climate change on air quality.

Major impedances to more detailed and quantitative evaluations of this type include our present inability to execute comprehensive global chemical models within the computational constraints that presently exist, and the coarse spatial meshes of the current GCMs. We can expect the first of these problems to be alleviated with faster computers and the coming generation of chemical codes. The second impedance—that of coarse GCM grid meshes—is particularly important because many of the meteorological phenomena in the above list occur on scales that are too small to be resolved within this structure. Many of the existing regional chemical codes could, for example, be adapted to provide useful air-quality predictions for global-change scenarios if only they were provided sufficiently resolved meteorological fields by the GCMs. A possible method to invoke may be the use of limited area models, which provide "physical interpolations" of GCM output (Giorigi et al., 1989). Since such GCM outputs will not be forthcoming for several years, one can expect correspondingly high levels of uncertainty in estimates of related air quality.

5.3.6 Fisheries

Although there have been a number of studies of the relationship between past climatic variations and marine fisheries (e.g., Southward et al., 1988), there have been fewer projections tied to global warming. According to Sibley and Strickland (1985), the distributions of most fish species are expected to move poleward, but the magnitude of the shift and the implications for yields are not known. The role of abiotic factors in affecting

The distributions of most fish species are expected to move poleward, but the magnitude of the shift and the implications for yields are not known.

the food supplies of immature fish is seen as a major source of uncertainty.

Recently, a number of papers have addressed global warming and freshwater fisheries. Meisner et al. (1988) describe how increased groundwater temperatures could affect survival and growth of salmonines (salmon, char, and trout) by altering temperature and dissolved oxygen in redds (nests). Areas having optimal summer conditions could shrink at low altitudes and latitudes and expand at high altitudes and latitudes.

The recent draft report to Congress by the Environmental Protection Agency (Smith and Tirpak, 1988) includes an analysis of possible effects of global warming on fisheries of the Great Lakes, California, and the Southeast. In the Great Lakes region, more favorable fish habitats are expected in fall, winter, and spring with enhanced productivity of open-water fish (bass, lake trout, and pike), which would more than offset less favorable summer conditions (decreased habitat, dissolved oxygen, and wetlands) that could harm fish populations. Both commercial and sport fishing would benefit. In California, salinity increases in the San Francisco Bay could enhance the abundance of marine fish species, while species breeding in low salinity and freshwater areas could be adversely affected. The species composition in subalpine lakes could be changed by higher temperatures and resulting increases in algal productivity. Negative effects were anticipated for coastal fisheries of the Gulf of Mexico, as finfish, shellfish, and crustaceans could be impacted by loss of coastal wetlands, temperatures above thermal tolerances, and increased salinity.

The prediction of climate change impact on fisheries (especially marine fisheries) is crude at this time, relative to more well-studied sectors such as agriculture, for several reasons: the physical changes in habitat (e.g., water temperature and circulation patterns) are less predictable at this time than are changes in air temperature and precipitation; fisheries, other than aquaculture, combine elements both of managed resources (i.e., economics, catch limits, and harvest technology are all important factors) and of unmanaged ecosystems; and basic life-history and population-dynamics information is often inadequate because of the vastness of the seas and sampling difficulty. It is also difficult to anticipate how abiotic conditions in inland waters will be affected by global climate change because it is not yet possible to relate the output from GCMs, with their coarse spatial resolution and crude approximation of surface hydrology, to the flow, water temperature, and water quality information needed for fisheries assessment.

A series of papers presented at a symposium on climate change and fisheries (Regier, 1988) explored a wide array of methodologies for forecasting the effects on marine and freshwater fish species and communities. The use of computer data bases on thermal tolerances of fish, bioenergetic models, early response indicator species, and large-scale experimental studies were among the forecasting approaches discussed. There was consensus that long and complex causal chains link climate change with ultimate effects on fisheries stocks (e.g., DeAngelis and Cushman, 1989). A combination of methodologies can provide a range of possible futures but cannot be viewed with confidence as a source of "predictions."

5.3.7 Coastal Zone

Early assessments of the impact of climate change on the coastal zone focused on

The kinds of resources at risk from rising sea level include both natural ecosystems, such as wetlands, and coastal structures. Increased salinization of ground- and surface-water supplies could be aggravated by precipitation and runoff decrease and threats to the operation of coastal sewage and drainage systems.

rising sea level, on the basis of relatively simple analyses of topographic data (Schneider and Chen, 1980, for the U.S.; Henderson-Sellers and McGuffie, 1986, for the world). The kinds of resources at risk from rising sea level include both natural ecosystems, such as wetlands (e.g., Armentano et al., 1986), and coastal structures (e.g., Kyper and Sorensen, 1985). Increased salinization of ground- and surface-water supplies (e.g., Hull and Titus, 1986) could be aggravated by decreases in precipitation and runoff and threats to the operation of coastal sewage and drainage systems (e.g., Titus et al., 1987; Wilcoxen, 1986). Coastal storms, such as tropical and extratropical cyclones and monsoons, can cause considerable damage to coastal structures and loss of life. The occurrence of tropical cyclones has been linked to sea surface temperature, which is expected to rise during a global warming; there is concern (with much uncertainty) that the frequency and severity of such storms could be increased as climate changed (e.g., Holland et al., 1987; Emanuel, 1987).

Although accounting for only 10% of the surface area of the ocean, the continental shelves provide over 90% of the worldwide fish catch (Falkowski, 1980). This disproportionate distribution of fish is related to the very high phytoplankton production that occurs on continental shelves. The phytoplankton form the base of the marine food chain. Over

A sea level rise of one meter by the year 2100 could inundate from 25 to 80% of the nation's coastal wetlands and cover about 7000 square miles of dry land.

the past 100,000 years the sea level has varied by over 100 m, and the siliceous and carbonaceous remains of phytoplankton in sedimentary rocks of the Great Plains are evidence of once higher production on relic continental shelves. A rising sea level could modestly increase the shelf area and concurrently increase the total amount of phytoplankton production globally. This could have two effects: (1) a reduction of the atmospheric CO_2 level as a result of increased phytoplankton photosynthesis; (2) an increased level of fish production.

The draft report to Congress by the Environmental Protection Agency (Smith and Tirpak, 1988) concludes that a sea level rise of one meter by the year 2100 could inundate from 25 to 80% of the nation's coastal wetlands and cover about 7000 square miles of dry land.[1] Possible responses to rising sea level, in developed areas, could include renourishment of beaches with sand pumped from offshore and construction of levees and bulkheads (with costs to coastal plant and animal communities whose habitat might be adversely affected). The cost-benefit evaluation of protective measures is unresolved, depending in part on the value assigned to wetlands.

Uncertainty limits our ability to predict the magnitude and rate of rising sea level from global warming. In particular, it is

[1] Even without climate change, many of these areas are susceptible to human activities and natural processes. More recent projections suggest that future sea level rise may be less than previously estimated (e.g., because of increased snow accumulation on the Antarctic and Greenland ice sheets).

known that coastal wetlands (Stevenson et al., 1986) and coral (Grigg and Epp, 1989) can accrete vertically and can keep up with rising sea level if the rise is not too rapid; beyond a critical rate, these systems would be inundated and destroyed. In terms of engineered systems, sea level rise can be accommodated if not too rapid because location, planned lifetime, and maintenance of coastal structures can be adjusted (National Research Council, 1987). However, a variety of coastal processes (such as tides, storm surges, and sediment transport, and coastal ecosystem dynamics) must be better understood before it is possible to predict the response of the coast to rising sea level (Mehta and Cushman, 1989). For example, in recent decades 28% of the mid-Atlantic coast has actually been accreting, rather than eroding (Dolan et al., 1989). Gornitz and Kanciruk (1989) have described an ongoing project to identify those areas of the world at highest risk from rising sea level based on the study of a combination of variables, including geology, geomorphology, tidal range, wave height, elevation, and regional trends in relative sea level and erosion/accretion.

5.3.8 Infrastructure

No credible estimates exist for the overall impact on infrastructure associated with a climate change. Issues surrounding the provision of infrastructure can be divided into three categories. First, there may be an existing infrastructure that will simply have to be moved; e.g., it may be in the way of rising seas, or it may be in place to service certain markets and populations that will migrate in response to changing climate. Second, there may exist an infrastructure that will have to be modified in some way or maintained differently; e.g., storm drains may have to be enlarged or river channels dredged more frequently. Third, new types of infrastructure

Infrastructure may have to be designed more flexibly than in the past to anticipate and reduce social, economic, and political displacements that could result from changing climate.

may be required—some based on technologies that already exist (e.g., new canals, new bulkheads) and some based on technologies that are yet to be developed (e.g., new transportation techniques, new energy sources).

In all of these cases, infrastructure may have to be designed more flexibly than in the past to anticipate and reduce social, economic, and political displacements that could result from changing climate. Long-lived capital investments will require creative reaction well in advance of climate change and this will be possible only if we develop decision-making structures that can deal with uncertainty and long time horizons. A new understanding of decision processes concerning long-term decisions is thus required, supported by research directed toward identifying the potential range of infrastructure stresses resulting from climatic and other changes.

5.3.9 Health Effects

Much of what is known concerning how climate change might affect human health has been inferred from correlation of health conditions with weather variables or seasonality. Recent studies that have focused on the possible impact that changing climate, season, and weather variables might have on the incidence of disease include: White and Hertz-Picciotto (1985), Wiseman and Longstreth (1988), Haile (1988), Kalkstein (1988), and Smith and Tirpak (1988).

Clear links have not yet been established between climate change and human health. Probably modest effects on human health,

Clear links have not yet been established between climate change and human health...Predictions of the health effects cannot be made without good predictive data on local temperatures, humidities, and levels of precipitation.

however, could occur through: (1) the direct impact of temperature (heat stress and cardio- and cerebrovascular conditions related to both summer and winter temperature extremes); (2) climate-related chronic, contagious, allergic, and vector-borne diseases (e.g., influenza and pneumonia, linked to the winter seasons; asthma and hay fever, linked to plants or fungi whose ranges and life cycles are strongly affected by climate and weather; and mosquito- and tick-borne diseases, such as encephalitis and Lyme disease); (3) premature birth, which has an adverse effect on human reproduction; (4) pulmonary conditions such as bronchitis and asthma related to urban and rural smog that may increase with climate change (Raloff, 1989); and effects of increased ultraviolet radiation on suppression of the immune system.

Other possible effects are somewhat more speculative. Climate-induced effects on agriculture, fisheries, water and coastal resources, and social and economic conditions might also affect human health. Decreases in food production might result in poorer diets, and rise in sea level and changed precipitation patterns may result in the deterioration of water supplies. Greater numbers of humans could migrate from one area to another, changing the geographic ranges and susceptibility of human populations to many diseases. In general, any event that reduces standards of living will have an adverse impact on human health (Chappie and Lave, 1982).

A number of issues have yet to be resolved. Predictions of the health effects cannot be made without good predictive data on local temperatures, humidities, and levels of precipitation. Confounding factors, some much more important than weather, affect human health. The relationship of these factors both with weather and with each other is complex, and, in many instances, stratospheric ozone depletion and global climate change might affect two or more factors simultaneously. We do not have the information required to accurately assess all the synergistic and offsetting effects.

Finally, we do not have much information on the social and economic impact that climate- or ozone-induced changes in mortality and morbidity might, in turn, generate. Nor do we have much information on the social and economic costs of such impact. Information on the out-of-pocket medical costs and the productivity losses associated with increases in morbidity, in particular, is not readily available.

5.3.10 Unmanaged Terrestrial Ecosystems

Understanding of terrestrial ecosystems is crucial to our ability to predict the regional- and global-scale consequences of potential greenhouse warming (e.g., Mooney et al., 1987). Climatic effects on terrestrial ecosystems are included here because of their functional role in global processes, their source of biotic and genetic diversity, and their role in the development and use of other natural resources as discussed above.

Refinements in simulation of forest growth have highlighted complex behavior in ecosystem response to climate warming: productivity and biomass depend on soil water availability, and positive feedbacks are evident regarding nutrient cycling (Pastor and Post, 1988). When climate-induced disturbances (e.g., fires) were modeled with forest growth, significant changes in biomass and composition were found at much higher rates than that due simply to climate warming alone (Overpeck et al., 1990).

Further potential global warming impact has been identified in a study based on climate-induced shifts of vegetative life-zones worldwide (Emanuel et al., 1985b). This analysis indicated that the largest changes might appear at high latitudes where the GCM simulated temperature increase was largest. Boreal forest zones could be replaced by either cool termperate forest or steppe; changes in the tropics were smaller. Also, tundra might be eliminated, and some expansion of grasslands and deserts was indicated.

Ecological studies of climate and vegetation in other biomes (e.g., arid-land ecosystems) also show a close link between atmospheric processes and patterns of community composition and structure (Neilson, 1986; 1987). A cause-effect hypothesis of recent desertification in the North American Southwest has been suggested in one study (Neilson 1986). Understanding of arid land processes is especially important because more than 25% of the Earth's surface is classed as arid or semi-arid, and these systems are extremely sensitive to abuse (Adams et al., 1978). The self-cleansing of pollutants that occurs in more humid regions is also nearly absent in arid regions, thereby exacerbating the problem of desertification (Dregne, 1983).

Examination of the fossil record suggests that both climate and vegetation have changed over the last 32,000 years (Woodward, 1987); such changes were probably worldwide, and data indicate that climate was the controlling factor. Other paleoecological studies (e.g., Davis and Botkin, 1985; Prentice, 1986; Webb, 1986; Davis 1989a,b) also provide support for at least the general predictions of forest-growth simulations, which indicate that changes in the distribution of individual species will modify terrestrial ecosystems significantly as a result of potential climate warming.

Because of the limited size and fragmentation of many natural ecosystems, concern is growing about the impact of global greenhouse warming on biodiversity of plants and animals. Biological reserves may lack the necessary size and migration corridors to facilitate dispersal as habitats change with progressive warming.

The effects of potential greenhouse warming on freshwater communities and aquatic biogeochemical cycles are unknown at present. However, freshwater ecosystems are integrators of watershed condition because of strong linkages (energy, nutrients, water) with their terrestrial setting (e.g., Likens, 1985; Minshall et al., 1985). Therefore, the impact of climatic warming identified above for terrestrial vegetation could translate to pronounced effects for freshwater ecosystems, given documented structural and functional responses in different biomes (e.g., Minshall et al., 1983).

The need to preserve biotic diversity of unmanaged terrestrial ecosystems has been well documented (Wilson, 1988). However, because of the limited size and fragmentation of many natural ecosystems, concern is growing about the impact of global greenhouse warming on biodiversity of plants and animals (Peters and Darling, 1985; Graham, 1988). Biological reserves may lack the necessary size and migration corridors to facilitate dispersal as habitats change with progressive warming. Specialized species, poor dispersers, and alpine and arctic communities represent examples of ecological resources at risk (Peters and Darling, 1985).

Prediction of the effects of climate change on terrestrial ecosystems is confounded by a number of uncertainties. For example, GCMs poorly predict local changes in precipitation, an important driving variable for ecosystems (Gates, 1985b), and little attention has been

Carbon dioxide, with its effect on leaf photosynthetic rate, transpiration, and water use efficiency, could enhance biotic storage of global carbon. Nevertheless, how leaf processes translate to ecosystem productivity and how CO_2 enrichment may affect decomposition is still largely unknown.

given in the models to ecosystem interactions (e.g., lack of vegetation over large areas) with a changing climate (Mooney et al., 1987).

Also, ecosystem disturbances from extreme events (which are also poorly predicted by GCMs) such as drought and large-scale thunderstorms (plus resultant effects such as wildfires) are known to shape landscape dynamics and heterogeneity of ecosystems (Pickett and White, 1985). A recent study indicates that global warming could favor such events, accelerate vegetative change, and alter the biomass and compositional response of forests to future warming (Overpeck et al., 1989).

The potential for biotic feedback mechanisms with climate change also confounds our understanding of how greenhouse warming might affect unmanaged ecosystems. Carbon dioxide, with its effect on leaf photosynthetic rate, transpiration, and water use efficiency (see Carlson and Bazzaz, 1980; Lemon, 1983; Martin et al., 1989), could enhance biotic storage of global carbon. Nevertheless, how leaf processes translate to ecosystem productivity (Gates, 1985a) and how CO_2 enrichment may affect decomposition (Mooney et al., 1987) is still largely unknown.

A final complicating factor is that potential global warming may take place within the context of other envieronmental disturbances. The impact of atmospheric pollutants (McLaughlin, 1985), acidic deposition on an international scale (Rodhe, 1989), and major changes in land use worldwide (Emanuel et al., 1985a; National Research Council, 1988) will contribute to uncertainties in our assessment of ecosystem effects from proposed climate change.

5.4 *Findings from Regional Studies*

A number of studies have attempted to produce an overall assessment of significant aspects of climate change on regional economies and societies. This section reviews the principal U.S. and international studies that have discussed the impacts of climate change, together with their results and major limitations.

5.4.1 *United States*

In 1988, the U.S. Environmental Protection Agency issued a draft Report to Congress entitled "The Potential Effects of Global Climate Change on the U.S." (Smith and Tirpak, 1988). That report focused on four regions of the U.S.: California, the Great Lakes, the Southeast, and the Great Plains. In each region, separate (and somewhat independent) studies examined the response of different sectors to climate change. Common climate projections were taken from the GCMs of the Geophysical Fluid Dynamics Laboratory, Goddard Institute for Space Studies, and Oregon State University; sea level rises of 50 to 200 cm by the year 2100 were assumed.

The following summaries of findings are taken from the draft Report to Congress:

- "Global warming could cause higher winter and lower spring runoff in California ... increasing the difficulty of meeting water supply needs. It could also increase the salinity in San Francisco Bay and the Delta and the relative abundance of marine species in the Bay; degrade water quality in alpine lakes; raise ambient ozone levels; and increase electricity demand. Changes in agriculture are uncertain."

- "Global climate change could affect the Great Lakes by lowering lake levels, reducing the ice cover, degrading water quality in rivers and shallow areas of the lakes. It could also expand agriculture in the north, change forest composition, decrease regional forest productivity in some areas, increase open water fish productivity, and alter energy demand and supply."
- "Global climate change could affect the Southeast by causing forests to shift to grasslands, reducing agricultural productivity and increasing the abandonment of farms, diminishing fish and shellfish populations, and increasing electricity demand higher than the national average. Approximately 90 percent of the national coastal wetland loss and two-thirds of the national shoreline protection costs due to climate change could occur in the Southeast. The impact on rivers and water supplies are uncertain."
- "Global warming in the Great Plains may reduce agricultural output, increase irrigation demand, change water quality, and increase electricity needs."

The "state of the art" in climate impact assessment has some significant methodological limitations (see Section 5.6 of this report). One limitation of special significance to the regional studies in the EPA draft Report to Congress is that the regional interactions between sectors are not fully accounted for. For example, in the California regional study, the effects of changes in agriculture on water resources and of changes in water costs on agriculture are not considered. Therefore, the findings of this report should be interpreted as a statement of the possible responses or vulnerability of some resource sectors to climate change. Ideally, a regional study should incorporate the many interactions and feedbacks that link the various sectors within the region.

Other regions of the U.S. have been examined in studies of particular resource sectors. For example, Flaschka et al. (1987) studied the response to climate change of the Great Basin, and Idso and Brazel (1984) and Callaway and Currie (1985) studied basins in the Lower Colorado region. Linder and Gibbs (1986) analyzed how electric utilities in New York State could be affected by climate change. In no case has a complete and integrated regional study been done. The first attempt to perform such a study is outlined in Cushman et al. (1989).

5.4.2 Regional Studies Outside of the United States

Potential effects of global climate warming also have been studied outside of the U.S., most particularly in the Netherlands, Canada, and the U.S.S.R. The impact on agriculture, forestry, and sea level rise has received the most attention. European and North American sites provide the greatest experience in dealing with historical sea level rise (Robin, 1986). Foreign locations that might be seriously affected by the predicted sea level rise of up to about one meter in the next 100 years include the Netherlands, Bangladesh, the mouth of the Nile River in Egypt, coastal areas of Japan, the coastal mangrove swamps of Indonesia and Malaysia, the area of Thailand around Bangkok, the Vilan Plain in northeast Taiwan, and many island atolls (Hekstra, 1989). Calculations have been done in a few instances of the probable cost to society. As an example, where structural adjustments to dikes, harbors, etc. are possible, a one meter increase in sea level in the Netherlands would cost about 10 billion guilders ($5 billion U.S.) over about 100

Foreign locations that might be seriously affected by the predicted sea level rise of up to about one meter in the next 100 years include the Netherlands, Bangladesh, the mouth of the Nile River in Egypt, coastal areas of Japan, the coastal mangrove swamps of Indonesia and Malaysia, the area of Thailand around Bangkok, the Vilan Plain in northeast Taiwan, and many island atolls.

years (de Ronde, 1989); for areas such as islands, however, abandonment may be the only option. Major uncertainties in foreign studies concern the impact on the ecological systems in estuaries and at coastal margins, the rate of net sea level change, and the potential rate of increase in sea level after the year 2050 (Hekstra, 1989).

A worldwide case study of climate variations on agriculture by the International Institute for Applied Systems Analysis (IIASA, see Parry et al., 1988a, 1988b) also included consideration of effects on timber markets. The study demonstrated economic benefits to consumers of wood products from increased boreal timber supply and economic damage to regions experiencing small increases in production (Sweden) having marginal production, or having made large investments growing trees (Brazil, Chile, New Zealand). Absolute values of the results were considered highly speculative owing to climate, biological, and economic uncertainties as well as long forecast periods. More detailed study of the forest products industry and possible mitigation measures are required.

Agriculture has been a major area of emphasis in international studies. The IIASA project pursued eleven agricultural case studies in temperate, high-latitude, and semiarid regions (Parry et al., 1988), some of which were integrated into their studies of regional economies. The case studies examined the impact of variations in climate using both climate anomalies (e.g., historical single-year droughts or historical series of several cold or warm years) and synthetic estimates of climate warming from GCMs. In general, the case studies of cold-limited regions such as Saskatchewan, Iceland, northern agricultural areas of the U.S.S.R., Finland, and Hokkaido in Japan showed some benefits from warmer growing season weather for agriculture. If, however, precipitation also decreases, production in some of these areas (such as southwest Saskatchewan) would be damaged by lack of moisture, moisture deficits over several years, or moisture at the wrong time. In general, the IIASA studies of semi-arid regions showed that interannual and intraseasonal variability in precipitation is an extremely important factor in the productivity of agriculture, both in subsistence agricultural areas such as Kenya, northeast Brazil, and Ecuador, and in commercial agricultural areas such as the southern U.S.S.R. and Australia.

The Canadian Climate Centre of Environment Canada recently published a series of regional studies of the effects of climate change on various individual resource sectors in the Canadian economy, together with an assessment of sector interactions in Ontario (DPA, 1988) and multiple resources in Quebec (Singh, 1988). Climate change forecasts for CO_2 doubling from the GISS GCM were used to drive the analyses. Almost all components of the climate system and resource use in the province were affected, including municipal water use, hydroelectric power, tourism and recreation, food production, forest resources, health, and residential heating and cooling requirements. In Quebec, for example, water supply to the James Bay area was projected to increase by 7% to 20%, while heating degree-days would fall by 25% in Montreal and by 35% in Quebec City.

The agricultural sector would see increases in growing seasons ranging between 22 and 72 days, depending upon the scenario. The forestry sector would experience a loss of boreal softwood forest of around 20% but an increase of 200% in hardwood acreage. Overall, the Ontario studies were less optimistic, pointing out, for example, a decrease in net basin supply of water to the Great Lakes on the order of 15%, which would in turn negatively impact water quality, tourism, hydroelectric power, and navigation on the Great Lakes (the latter would be somewhat offset by increased length of the ice-free shipping season). The Canadian studies also examined agricultural impact in Ontario and the prairie provinces, sea level rise in the maritime provinces, and the skiing industry in Ontario.

5.5 *Findings from Analog Studies*

A number of historical studies have been conducted to learn the effects and responses of global climate change on human and natural systems. Early studies have provided assessments of the impact of climate on human societies, for example Wigley et al. (1981); Rotberg and Rabb (1981); Lamb (1982). A fairly lengthy list of early analog climate impact studies is provided by Kates et al., (1985). Properly qualified, these "semidescriptive" case studies may shed useful light on the relationships between society and climate change. Some of the most thorough and interesting recent assessments have been provided by Glantz and his coauthors (Glantz and Krenz, 1987; Glantz, 1988) and by the IIASA study on the impact of climate variations on agriculture (Parry et al., 1988a, 1988b).

Glantz (1988) provides a number of case studies of the responses of U.S. and other institutions to environmental changes similar to those expected from future CO_2-induced climate change, including areas as diverse as

Case studies have demonstrated how societies can respond to climate variability. For example, coalition-building among governments and interest groups has proven to be important if action is to be taken in the face of uncertainty.

fluctuations in the level of the Great Salt Lake and Great Lakes; decline in the flow of the Colorado River; drought in California; repeated freeze periods in citrus agriculture in Florida; sea level rise in Charleston, South Carolina and Louisiana; and variations in the navigability of the Mississippi River. Parry et al. (1988a, 1988b) cover foreign societal responses to past historical climate-related fluctuations in agriculture. This includes relatively successful risk-minimizing cropping patterns and other institutional adaptations of indigenous peoples to climate variability of the Ecuadorian Andes, the successful food purchase programs of the Kenyan government in response to drought, and the reliance of Icelandic farmers on fodder reserves to guard against cold summers. In some cases, ad hoc or emergency responses have themselves been institutionalized to the point that the population has come to expect them, as in India (Gadgil et al., 1988). In other cases, such as the Colorado River Compact, the institutions are rigid to the point that surface water is used to grow low-value crops, and cities in the region are mining groundwater at a high cost (Brown, 1988).

In general, case studies have demonstrated how societies can respond to climate variability. For example, coalition-building among governments and interest groups has proven to be important if action is to be taken in the face of uncertainty. All of the case studies raise the issue of intergenerational equity. The studies show that ad hoc responses and traditional approaches usually have been the preferred initial mode of adaptation, which in turn often added an element of rigidity to further response. Some of the studies show that once the regional winners and losers from environmental change have been identified, the winners have little incentive to compensate the losers. The analog studies contain a wealth of detail, integrate a broad range of knowledge, provide a multiplicity of perspectives, and are easy to communicate and use. The disadvantages of analog studies include lack of definite causes for environmental change that can be related to greenhouse warming (the cause is not always known) and the simple failure of the analogy to be appropriate for a warmer world with an elevated CO_2 concentration (many of the studies concern cooling, and none has elevated CO_2).

5.6 Uncertainties and Information Needs

This section deals with limitations on effects estimation that are posed by uncertainty or lack of information concerning biophysical and socioeconomic relationships, models, and data required to estimate the effects of climate change on the human environment. Both limitations in existing methodologies and problems with available data are addressed.[2]

5.6.1 Limitations in Methodology

Climate problems dominated by enormous uncertainty have, in the past, been handled by sensitivity analysis (Smith and Tirpak, 1988) and Monte Carlo simulation techniques (Edmonds and Reilly, 1985; Reilly et al., 1987; Nordhaus and Yohe, 1983). Structural uncertainty has usually been handled by examining the outputs of alternative model structures (Bolin et al., 1986). The climate change system related to biophysical and socioeconomic effects is, however, so complex that "brute force" application of these techniques is not likely to be productive in providing significant insight into the important causes of certain effects or the value of information in foreseeing their potential ranges. New methods based on sound statistical theory are thus required to sort through the myriad relevant variables if the state of the art is to be advanced in application to climate change effects (see Liebetrau et al., 1990, for first steps in that effort).

Several criteria can be used to accomplish this sorting. Potential response variables should be included (Brainard, 1967). It may, however, be possible to identify sets of collinear variables that would allow researchers to focus their attention on only one of the variables as they construct their models (Fair, 1980; Box et al., 1987). Input designs to meet specific modeling objectives are being developed and may prove useful for sorting or ranking input variables (Sacks et al., 1989). Uncertainty about some other variables may not be important even though their future trajectories are not well known; median or mean values of their trajectories can thus be assumed with little cost. It can be expected that any model that provides access to social and economic effects will be composed of many submodules, and the relative importance of these modules can be critical. The potential range of values for variables that have a small effect on the output of an important module might be investigated thoroughly, and the range of a variable that has a large effect on the output of a relatively unimportant module might be given low priority. A new methodology based on hierarchical modeling that systematically sorts on the basis of these and other criteria is currently under development as part of the DOE

[2] Risk assessment approaches are not invoked here because scenario analyses are to be conducted as a separate aspect of the development of the National Energy Strategy.

program described in Cushman et al. (1989). Similar methodologies are being developed in other areas that may be useful in the global modeling arena (Chapter 6 of Reimus et al., 1989 for geological storage of nuclear waste and Gilbert et al., in preparation, for an application to radioactive dose estimation).

5.6.2 Limitations in Information

Environmental assessments of the impact of climate change have focused on individual resource sectors in selected regions of the U.S. A major reason is that regional studies require a major commitment of time and money; a truly integrated assessment requires a very significant commitment. A recent example is the National Acid Precipitation Assessment Program (NAPAP, 1989). This section focuses on the general limitations in information for environmental assessments at the regional and national scales; detailed lists of data needs have been prepared for individual resources (White, 1985). Numerically and geographically integrated data sets for environmental data do not exist for large regions (Olson, 1984)—let alone associated economic data sets. Economic and social data are usually collected and reported by political unit (county, census district, state) whereas data on natural resources are often collected and reported by habitat-type or by physically defined areas such as watershed or soil map units. No standard methods exist for integrating data collected for different spatial units or for defining study regions to minimize uncertainties caused by boundary heterogeneity. Large, integrated data base systems that provide data stored in compatible spatial and temporal formats, with associated analysis and mapping capabilities to conduct integrated studies, are rare.

This is partially true because the U.S. lacks a history of consistent regional and national data acquisition planning between federal and state agencies. Also, long-term maintenance of integrated data bases is difficult because of funding cycles and changes in perceived need. Examples of existing integrated data systems include the Department of Energy's GEOECOLOGY and SEEDIS, the Council of Environmental Quality's UPGRADE, the Environmental Protection Agency's GEMS and, with the Department of Energy, ADDNET, and DAT GRAF (Merrill, 1982; Olson et al., 1987). Typically, one third to one half of the effort spent on regional studies is devoted to developing an integrated and quality-assured data base.

Although regional studies have been performed for many years, the ecosystem properties that are important for regional scales are still poorly understood (Hunsaker et al., 1989). Few regional-scale biological models exist. In most instances, either local models will have to be scaled up, which can introduce many problems and uncertainties (Solomon, 1986), or entirely new models will have to be built (Emanuel et al., 1985b). The number of available models for physical processes can also be limited. For example, the models are not available to adequately assess the impact of climate change on water resources at the local, regional, and national scales. Resource and economic models with input and output parameters that facilitate the linking of several resource models for the assessment of regional impacts have not been developed.

Data manipulation and extrapolation can contribute to uncertainty because of inadequate spatial or temporal resolution, or both.

No standard methods exist for integrating data collected for different spatial units or for defining study regions to minimize uncertainties caused by boundary heterogeneity. . .the U.S. lacks a history of consistent regional and national data acquisition planning between federal and state agencies.

Point data for large geographic regions are often uneven in quality and distribution. Often it is difficult to find time series data for the same period of record for several environmental parameters. This can be especially limiting if the analog approach to assessment of impact is being used. Some data such as land use and soil chemistry are available for only a few points in time. Computerized, digital terrain data of sufficient spatial resolution to be useful for coastal studies (such as land inundation from a sea level rise of 1 m) are only available for limited areas (Durfee et al., 1986; Bright et al., 1988). The classification of geographic areas according to the relative homogeneity of one or more environmental attributes can be useful in reducing uncertainty if the classification scale is appropriate to the disturbance; however, the contribution to assessment uncertainty from such classification needs further investigation. Research is needed to assess the effect of landscape patterns on regional assessments. Some of the more recent technological tools—such as geographic information systems, satellite sensors that capture biologically significant spectral patterns, and supercomputers that can process large spatial arrays—will be useful for addressing the theoretical and applied research challenges that the regional scale poses (Hunsaker et al., 1989).

The goal of a current DOE study (Cushman et al., 1989) is to show how an integrated regional assessment of climate change can be performed. During the first year of the study, data sets and models are being developed for agriculture, water use and supply, forest resources, and economic factors. A goal of the second phase will be integration across natural resource models. Treatment of feedbacks and couplings on scales of decades to centuries remains to be addressed.

Understanding of the potential effects of climate change is based on subjective views of the likely distributions of myriad random variables. It is therefore critically dependent upon our ability to foresee future developments.

5.6.3 Dependence on Forecasting of the Future

Understanding of the potential effects of climate change is based on subjective views of the likely distributions of myriad random variables. It is therefore critically dependent upon our ability to foresee future developments that will affect both temporal changes in these distributions and the "general equilibrium" interactions among them, which work across the entire system. Bolin et al. (1986) observe that no existing model could have predicted the development of the global energy system in the 1970s using data available in the 1950s and 1960s; simple statistical extrapolation of price data from those years does, however, capture the 1973 and 1979 oil price shocks in the 80% confidence interval (Glynn and Manne, 1988). We need to understand that learning process as well as we understand the system itself.

Some of our learning will be Bayesian in nature—simply improving our understanding of possible ranges of critical state variables by observing how they move into the future. We will learn more quickly about some variables than others, and we will need to assess the value of resources devoted to the process, weighing the relative ease of learning about each variable against its relative importance in affecting human existence. Other learning will be derived by improved understanding of underlying driving variables and the processes by which they affect the critical state variables. Recognizing, as a result, that the subjective distributions of state variables are really conditional distributions, we must also assess the value of resources devoted to investigating these variables and interactions. Still more learning will occur as we discover entirely new interactions and uncertainties. This sort of new insight can be expected to expand the base of knowledge with which we view the future and thereby increase the value of the other types of learning. Investigation into all of these processes in the context of the type of long-term uncertainty of climate change effects is under way as part of the DOE effort described in Cushman et al. (1989).

5.7 Research Needs

The state of science with respect to the effects of global climate change on the human environment is still in its infancy. While considerable work has been done on the impact of various steady-state climate change scenarios and historical weather on selected natural resources, the resultant economic and social consequences have not been thoroughly examined. In general, research has revealed probable qualitative effects for selected historical and synthetic climate patterns thought to be illustrative of future climate change. However, projecting the quantitative effects of climate change remains an elusive goal for the moment.

The necessary input to a quantitative analysis of the effects of CO_2/climate change is output of a long sequence of other projections including greenhouse gas emissions, the disposition of greenhouse gases, and time-dependent scenarios of regional climate change, not to mention projections of societal and economic responses to these changes. Researchers cannot hope to project CO_2/climate change impact without accurate projections of CO_2/climate change effects. Even if such CO_2/climate projections were available, a vast research effort mounted over the course of years would be required before adequate projections of impact could be made.

Many of the uncertainties in the area of effects on human society are still at the conceptual or process level and are site-specific.

There are methodological and data-related problems with respect to quantitatively characterizing the complex underlying interactions between man and his environment. More important, the current capability of climate models to replicate current climate is limited to continental-scale effects, while events important for impact on humans occur at much smaller local and regional levels.

Table 5.1 summarizes the general state of the science or "knowns" as well as the significant uncertainties. Unlike the specific rates of emission for certain of the greenhouse gases or specific parameters in equations for GCMs, where a range of values sometimes can be stated to quantify the degree of uncertainty, many of the uncertainties in the area of effects on human society are still at the conceptual or process level and are site-specific. For example, in the area of agriculture, where some of the most extensive work has been done, only the basic outlines of models of plant physiology and their relationship to CO_2/climate change have been constructed. It is not yet clear how significant the timing of weather events and CO_2 fertilization will be. In addition, it is not yet clear what the reactions of the agricultural sector will be (e.g. choice of crops, irrigation and pesticide and herbicide use, planting dates, etc.), though they will likely have a major influence on the magnitude of the effects. The lack of detailed regional climate forecasts is a major source of uncertainty; nevertheless, there is still major uncertainty that needs to be resolved within the models themselves.

Table 5.1 Summary of state of the science and uncertainties in effects of global climate change on the human environment.

Parameter or area of inquiry	Knowns	Uncertainties
Findings of sectoral studies		
1. Energy	Probable general direction of effects on conventional energy supply (except hydro) and demand, given a level of average temperature increase. Models and methods for forecasting short-term weather variations.	Regional effects of climate change on weather variables. Effect of climate changes on hydroelectric supply. Effect of climate change on biomass supply, wind energy, and the productivity of ocean thermal and other unconventional resources. Socioeconomic and biogeochemical feedbacks.
2. Agriculture	Mechanisms of climate change. Physical crop models with some CO_2 fertilization from laboratory experiments.	Necessary detailed regional climate forecasts are very uncertain. Farmer responses to weather under changed climate are not clear. Because of these and other considerations, even the direction of effects on specific crops in given locations is uncertain.
3. Water resources	Models of watersheds, ground water supply, and some river basins. Relationship of precipitation to runoff and water supply for today's climate.	Necessary detailed regional and temporal forecasts of temperature, precipitation, and other weather variables. Relationships between small-area precipitation and large-basin water supply. Effects of changed seasonality of precipitation. Human institutional response.
4. Forestry	Mechanisms of climate impact. Physical models of forest succession for small plots. Estimates and models of world wood products markets.	Regional weather input for forest succession models. Existence of a CO_2 fertilization effect. Response of the forest industry to high rates of climate change.

Table 5.1 Summary of state of the science and uncertainties in effects of global climate change on the human environment (continued).

Parameter or area of inquiry	Knowns	Uncertainties
5. Air quality	General effects of temperature on severity of inversion episodes.	Likelihood of inversions (requires weather forecasts). Synergistic and offsetting effects of pollutant emissions.
6. Fisheries	General movement of marine fisheries poleward. General influence of temperatures on freshwater species.	Quantitative influence of warming on current and abiotic processes. Influence of warming, precipitation, and runoff on freshwater habitats.
7. Coastal zone	Probable rates of sea level rise in many loations, given a scenario of temperature change. Costs of coastal defense.	Impact on, and value of, coastal wetland resources. Effects on land subsidence and coastal fresh ground water. Effects of and on coastal processess such as sediment transport.
8. Infrastructure	Some data on the influence of weather phenomena on road and building maintenance requirements, utility demand, hydroelectric supply, irrigation works. Influence of sea level on coastal infrastructure.	Relationship of global warming to local weather. Quantitative estimates of infrastructure requirements.
9. Human health	Some inferential data on the relationship of health conditions to weather episodes, disease vectors, and climate.	Influence of global warming on proximate causes of disease and health conditions. Influence of synergistic and offsetting factors.

Table 5.1 Summary of state of the science and uncertainties in effects of global climate change on the human environment (continued).

Parameter or area of inquiry	Knowns	Uncertainties
Findings of regional studies	Impact of global warming scenarios (usually a CO_2 doubling) on individual regional resources, given current technology. Some discussion of intersectoral links and sensitivity of impacts to human adaptation.	Realistic warming scenarios. Scientific basis for resource effects (e.g., open-air CO_2 fertilization). Influence of human adaptation and technological change on impacts. Effects of realistic intersectoral and interregional economic and social linkages.
Findings of analog studies	Impact of institutional adaptation on environmental impacts. Reactions of human institutions to complex environmental problems with interregional, intersectoral, and intergenerational implications.	Appropriateness of the analogs studied to a global climate change situation.
10. Unmanaged terrestrial ecosystems	Forest response based on simulations. Some climate-induced vegetation change based on the fossil record.	Modeled response of other biomes. Interactions and feedbacks with biogeochemical cycles and vegetative composition. CO_2 enrichment effects. Lack of experimental and observational tests of predictions. Changes in terrestrial-acquatic linkages. Role of climate-induced disturbances like drought and wildfires.

For analytical convenience, these studies have focused on deterministic, long-run effects of climate change equivalent to a doubling of atmospheric CO_2 in the steady state, typically for a few resources in isolation. They have not addressed many of the key issues in predicting the effects of climate, which include:

- Time. Resource effects will not take place in today's world with today's technology, nor will the effects occur for the most part in a world that is steady state. Some human responses to climate change (particularly those easy to change) will be influenced at least partly by current climate conditions, not those that might eventually come to pass. Second, human response to the change in climate and its environmental effects will be determined in part by the technology available at the time—which is, in turn, a function of learning over time. The rate and timing of CO_2/climate change may be more important to determining consequences than the characteristics of a steady-state CO_2/climate.
- Simultaneous Multiple Resource Analysis. Methods must be developed to address the interactions of human and natural systems, including methods to simultaneously analyze the interaction of multiple resources within a systems framework, quantify the combined CO_2/climate change interactions, and forecast trends in exogenous system variables such as technological change and population growth. Because many economic agents in a region may compete for the same regional resource (e.g. surface water), the indirect market effects of climate change may either exacerbate or mitigate the direct effects. This implies that the analysis of impact in a region must integrate all the significant environmental, economic, and social interactions.

Future climate change will be experienced as day-to-day changes in weather-temperature, precipitation, wind speed, humidity, etc., and it is uncertain what the physical consequences of climate change will be.

- Geographic Disaggregation and Integration. To be meaningful, effects studies must select appropriate geographic entities for analysis (site, regional, global) and because of the interactions between regions, conduct simultaneous multiple region analysis. These effects occur through either interregional trade and market responses or human migration. A hierarchy of analysis and models spanning the geographic dimensionality from local to regional to global must be developed.
- Uncertainty. Studies must develop and apply appropriate uncertainty analysis techniques. Future climate change will be experienced as day-to-day changes in weather—temperature, precipitation, wind speed, humidity, etc., and it is uncertain what the physical consequences of climate change will be. Because we are also uncertain what the rate of increase in greenhouse gas emissions will be (causing uncertainty in the rate of temperature change), climate models are not yet accurate enough to predict regional changes in atmospheric behavior. Because future economic growth, populations, and technologies are also uncertain, analyses of the environmental consequences of climate change will have to explicitly indicate uncertainties and their effects on estimates of impact. It is likely that this work will require advances in analytical and computational techniques.

Chapter 6: OPTIONS FOR TECHNOLOGICAL RESPONSE TO CLIMATE CHANGE

Possibilities for human response to climate change generally revolve around a tripartite distinction between prevention, mitigation, and adaptation. Prevention is the strategy of encouraging changed patterns of energy use that, in turn, would contribute to reduced emissions of greenhouse gases. Examples include switching away from fossil-fuel use, using fossil fuels more efficiently, and removing greenhouse gases from point-source emissions. Mitigation generally is recognized as a category of options for reducing the impact of greenhouse gases on the environment. Proposals to increase the Earth's albedo to compensate for radiative heating due to increasing atmospheric concentrations would seek to counteract the heating effects without reducing greenhouse gas emissions. However, the category of mitigation is somewhat ambiguous. Sometimes it includes the removal of CO_2 from the atmosphere by planting biomass to offset emissions. However, as this would reduce the atmospheric concentration of CO_2, it might be construed by some as net emissions reduction and, therefore, a prevention strategy. Another mitigation measure might be the construction of sea defenses against coastal intrusion. The third policy strategy is that of allowing climate to change and then adapting human systems to the changing environment, and perhaps assisting natural systems to adapt also. Long-term actions to cope with sea level rise, such as retreat from the shoreline, also are adaptions.

To some extent, the distinction among prevention, mitigation, and adaptation represents a sequential range of possibilities depending on how advanced the process of global warming is at the time action is taken. However, the extremes of prevention and adaptation have become polar positions of policy advocates, each regarding the other as morally irresponsible for advocating either irreversible delay or precipitous action, respectively. This adds to the burden of ambiguity associated with its middle term, mitigation, to inspire us to seek a clearer set of options.

Four sets of options are examined: fix, change, tinker, and cope. These response options are intended not to represent either a sequence or a mutually exclusive choice of response options, but to represent a menu of possibilities from which a mixed strategy could be selected to produce the desired end of slowing the impact of global warming.

For these reasons, four sets of options are examined: fix, change, tinker, and cope. These response options are intended not to represent either a sequence or a mutually exclusive choice of response options but to represent a menu of possibilities from which a mixed strategy could be selected to produce the desired end of slowing the impact of global warming. For all options, incentives may be needed to encourage their implementation.

Fix

This category of options encompasses those designed to reduce the rate of greenhouse gas emissions without requiring a major shift away from fossil fuels. These options would include changing the fossil-fuel mix to expand the proportion of natural gas that is used. With or without fossil-fuel switching, increased energy efficiency on both the demand and supply sides could offer major opportunities for emission reduction, provided demand remains fairly inelastic. Economic or other restructuring to reduce demand for services that require energy is also an option in this category. Biomass offsets and cost effective measures for removing and sequestering carbon from point-source emissions would allow continued use, or even expanded use, of coal or other fossil fuels.

Change

Movement away from fossil-fuel use would certainly reduce the rate of CO_2 emissions. Obvious options include noncombustion technologies such as nuclear, hydroelectric, photovoltaic, wind, and geothermal technologies. However, efficient biomass combustion technologies (such as biomass gasification linked to combined cycle combustion) also hold out great promise. Although this technology does emit CO_2, it has the potential to stabilize the atmospheric concentration by closing the carbon fuel cycle in years rather than the millions of years required by fossil fuels.

Tinker

Techniques of climate modification designed to counteract directly the effects of greenhouse gases fall into this category. Such options include attempts to change the Earth's albedo through stratospheric injection of sulfate aerosols, altering radiative fluxes by increasing the albedo at the Earth's surface, and controlling feedback processes. Although perhaps somewhat audacious, examination of such schemes provides perspective for the magnitude of the greenhouse effect.

Cope

Corresponding closely to the established strategy of adaptation, this category consists of a wide range of options, including responding to sea level rise through sea defenses or shoreline retreat; changing patterns of water resource use in energy production, agriculture, and recreation in response to precipitation changes; and development of crop types suitable for changing growing seasons.

This chapter presents a selection from the options identified in the technical literature as possible technical responses to global warming. The range of ideas presented here is indicative of the concepts that have been offered—from the genuinely practical to the clearly fanciful. This summary should not, by any means, be accepted as all-inclusive. More detailed treatment of many related aspects of these options will be considered in companion reports prepared under the National Energy Strategy (e.g., fossil energy, alternative energy technologies, and the potential for conservation and energy efficiency). Much room remains for creative thinking.

6.1 Fix-Options for Reducing Emissions from Fossil-Fuel Systems

6.1.1 Increasing the Efficiency of Energy Conversion and Use

Improving the efficiency of energy use necessarily requires an understanding of what the energy is used for and how it is used, together with a careful tailoring of technology to these circumstances. Nevertheless, improvements in energy efficiency are possible in a broad set of categories.[1] Proven technologies exist to improve energy efficiency in each of these categories, although large improvements in some categories would require development of additional technology.

Improving the fuel economy of transportation vehicles, especially cars, trucks, and

[1] See also the companion multi-laboratory report on energy efficiency for a discussion considering more than aspects related to greenhouse gases.

Improving the efficiency of energy use necessarily requires an understanding of what the energy is used for and how it is used, together with a careful tailoring of technology to these circumstances.

commercial aircraft. Although new cars average slightly more than 28 miles per gallon (MPG) (U.S. DOE, 1988), prototype cars have been developed that get more than 60 MPG in city driving and more than 80 MPG on the highway. These prototypes use lighter materials to reduce vehicle weight, body designs that reduce aerodynamic drag, high-efficiency engines to improve extraction of energy from fuel, electronic or continuously-variable transmissions to reduce drive-train losses, and energy storage systems to recover energy lost in braking. Similar technological approaches can be used to reduce fuel consumption in trucks, and some of these could also reduce aircraft fuel consumption. Research and development on advanced engine designs, perhaps including ceramic diesel engines, could improve the fuel economy of highway vehicles even further. Prop-fan aircraft engines have been demonstrated that yield comparable performance to jet engines in commercial aircraft applications and use less energy; other engine designs, body designs, materials, and aircraft surfaces could, in combination, improve fuel economy by more than a third over the best of present designs, although at higher vehicle costs.

Requiring greater energy efficiency in buildings. The amount of heating and cooling that a building requires can be reduced by improving building shells by superinsulating, shading by trees and architectural design, using passive-solar features, using precision manufacturing systems in construction to reduce air infiltration, incorporating heat-cool storage in some climates, and using low-emissivity window glass. The first four of these technologies can be difficult or expensive to retrofit into existing structures. Using these five technologies, it is possible to build new buildings that use 50% or less energy than the average existing building in the U.S. requires for space conditioning. Combining the best of the available technologies could reduce energy consumption for space conditioning in new residences by 90% compared to average existing residences (Goldemberg et al., 1987). Other technologies can improve the efficiency of energy use within both new and existing buildings. These include high-efficiency space-conditioning systems; modular-energy systems that combine space-conditioning, water-heating, and electricity generation; and improved lighting and energy-management systems. Most of these technologies can be retrofit into existing structures and the efficiency gains could be significant. Gains may be at even less cost and higher efficiency in a new structure designed to use them efficiently.

Combining the best of the available technologies could reduce energy consumption for space conditioning in new residences by 90% compared to average existing residences.

Improving the efficiency of household and commercial appliances. In addition to the space-conditioning technologies listed above, commercially available lighting equipment can reduce the energy required for lighting by more than 70% in some commercial settings. Some commercially available refrigerators use additional insulation and improved compressor and motor designs to reduce electricity consumption by 50% from that of the average refrigerator. Technology exists to reduce energy consumption even further, although at higher equipment cost. Reducing the energy consumption by these and other appliances can reduce the air conditioning needed to remove the wasted energy.

Improving the heat rates of new and existing power plants. Improved maintenance and repair practices can yield modest reductions in the energy required to generate a kilowatt-hour of electricity. Repowering existing units, typically by adding a gas turbine cycle to a steam generating plant, is more expensive but can reduce fuel consumption and CO_2 emissions by 10% or more. Aeroderivative gas turbines, and some "clean coal" technologies being developed to reduce sulfur dioxide emissions, can reduce fuel consumption in new power plants by 10 to 20% when compared with existing plants.

Substituting DC adjustable-speed electric motors in many industrial, commercial, and residential applications. Large AC motors that operate at continuous loads are common in pumps, compressors, and fans. The potential for energy savings is significant because 70% of industrial electrical energy use is for motor drive. In many of these applications, the motor capacity is determined by the starting load, not by the lower needs of the average load, and the oversizing increases energy consumption. Substituting DC motors and power-electronics control systems can reduce electricity consumption during continuous operation by 15 to 20%.

Beyond these categories, for which general technologies and policies may be appropriate, there exist a large number of materials processing and manufacturing systems for which highly individualized applications of technology and policy can improve energy efficiency. Examples include reducing electricity use in aluminum, silicon, and steel production and gas use in petroleum refining.

All but the last of these general categories involve decisions about capital stock that may provide service for 10 to 40 years or more; not all uses of electric motors are so long-lived. Large markets exist for used buildings, vehicles, and equipment so that the purchase of a new piece of energy-efficient capital stock does not necessarily remove an inefficient piece from use. The energy efficiency of new buildings and equipment could be improved relatively quickly, and of new vehicles in perhaps five years, but these improvements would have little effect on aggregate energy consumption by buildings, equipment, and vehicles until the new investments become significant shares of total investment, in 10 to 20 years. Accelerating the replacement of existing inefficient stock can be very expensive, given the size of existing stocks.

Improving the efficiency of existing stock must overcome behavioral and institutional as well as economic obstacles. In all of the categories except improving power-plant heat rates and the fuel economy of commercial aircraft, energy costs are a small portion of the cost of purchasing, using, and maintaining the equipment. Even though improving energy efficiency may pay for itself, decision makers often consider the benefit to be too small to bother with, relative to improvements in other costs or measures of performance. This appears especially true of investments to improve existing equipment rather than to replace it. In addition, decision makers often behave as though they use discount rates of several hundred percent when calculating the future benefits of lower energy consumption against the present investment costs of purchasing efficient equipment. However, it should be borne in mind that the discount rate in such cases is a surrogate variable that may be loaded with factors such as perceived risk of the investment and the expected life span of the equipment in addition to the time-cost discount of the value of money. Careful packaging of information, financing, incentives, delivery of services to improve efficiency, regulation, and other options are necessary to overcome these obstacles for

Even though improving energy efficiency may pay for itself, decision makers often consider the benefit to be too small to bother with, relative to improvements in other costs or measures of performance ... Careful packaging of information, financing, incentives, delivery of services to improve efficiency, regulation, and other options are necessary to overcome these obstacles for different groups of decision makers.

different groups of decision makers (e.g., different social or economic groups of homeowners, different types or sizes of businesses), even though the basic technologies (e.g., insulation, more efficient appliances) are essentially similar for all users.

Some options for improving efficiency require changes in the behavior of homeowners, vehicle operators, and other decision makers. For example, vehicles and other equipment often consume more energy than their designers intended because of poor installation, operation, or maintenance; high driving speeds and failure to clean filters and condenser coils are probably the best known examples of behavior that can increase energy consumption. Equipment design can compensate for only some of this behavior. Changing this behavior may require combinations of higher energy prices, subsidy of maintenance costs, information, licensing, or regulations to encourage people to give up some of the convenience or speed they gain by not operating equipment efficiently. Measures that are effective may be costly or politically difficult to implement. Moreover, although changing behavior can be effective in the short term, more research is needed to understand how best to sustain such changes over longer periods of time.

It is often technically possible to improve energy efficiency to levels beyond what would

Within the realm of increasing the efficiency of energy use, there is no single item that can be a major contributor to reducing CO_2 emissions, but there are many possibilities that could have, collectively, a large impact.

be justified economically given today's energy prices, or even a doubling or tripling of these prices (whether by market forces or taxes). Implementation of such options might require additional research and development to reduce their cost or, with options such as very thick insulation where the potential for reducing costs appears limited, regulatory and fiscal incentives to require their use and offset the cost or other effects of their adoption.

Within the realm of increasing the efficiency of energy use, there is no single item that can be a major contributor to reducing CO_2 emissions, but there are many possibilities that could have, collectively, a large impact.

6.1.2 Fuel Switching Among Fossil Fuels

Largely because of differences in fuel chemistry, different fossil fuels have different rates of CO_2 emissions per unit of useful energy. This leaves open the possibility of reducing total CO_2 emissions by switching from coal to natural gas in applications where either fuel would be suitable. As an average, the combustion of natural gas releases 13.78 kgC per billion joules of energy, whereas liquid fuels release 19.94 kgC per billion joules and coal 24.12 kgC per billion joules. These are not, however, the numbers that should be used for considerations of fuel switching because they do not acknowledge the energy use (and CO_2 emissions) embodied in refined petroleum products, for example, as a result of exploration, development, refining, transportation, etc.

Although CO_2 savings may be achievable through fuel switching, care must be taken to insure that it is full-fuel-cycle CO_2 costs that are compared. For example, there may be counteracting effects of switching if other greenhouse gases are released. Because CH_4 is about 30 times as efficient a greenhouse gas as CO_2, the release of methane from coal mining and the leakage of natural gas must be considered in estimating the net greenhouse effect.

Over the long term, resource constraints will also play a role in these comparisons because natural gas resources are much more limited than are coal resources based on current estimates. This is a particularly relevant issue in the U.S. where energy security is a significant issue and the U.S. is blessed with very large coal resources. Because the total carbon content of the recoverable resources of conventional crude oil and natural gas are small in comparison to recoverable coal resources, the atmospheric concentration of carbon dioxide can only approach a level near double its preindustrial value if there is utilization of large quantities of coal.

The importance of considering full-fuel-cycle CO_2 costs is central to evaluation of synthetic gaseous and liquid fuels from coal. Although CO_2 emissions may be no different at the point of final combustion, there are generally large fuel-usage and conversion losses (and CO_2 emissions) during the conversion process.

Utilities already are interested in some high-efficiency generating technologies for reasons unrelated to greenhouse gas emissions. For example, the Department of Energy is supporting the development of new clean coal technologies designed to reduce the emission precursors of acid precipitation. Some of these technologies are anticipated to use fuel more efficiently than present coal combustion technologies. To the extent that they replace existing coal-fired capacity, these technologies have the potential to reduce CO_2 emissions. Other types of clean coal technology, however, may result in a net decrease in the conversion efficiency of coal to electricity, hence requiring the burning of more coal to achieve the same output.

International experience ... has demonstrated that significant fuel switching in transportation can be accomplished but that it requires a concerted effort.

Fuel choice in some applications, particularly power generation and some industrial processes, is very sensitive to the relative prices of alternative fuels. Policies to reduce the cost of natural gas relative to that of petroleum and coal would encourage the substitution of this fuel in these applications. Policies that require control of other types of emissions from coal, such as SO_2, would have the effect of increasing the cost of using coal relative to that of using gas and could help to reduce the degree of taxation or other intervention in energy pricing needed to effect fuel switching.

In some applications, especially in transportation, end-use technology and the institutions for supplying it have developed to use only one fuel. Although technology exists to use ethanol, methanol, methane, and hydrogen in transportation, introduction of the fuel and of the end-use technology requires coordination among different groups of decision makers. Fuel suppliers require assurance that if they begin investing now in order to supply an alternate fuel three to five years from now, vehicle manufacturers will supply vehicles that can use the alternate fuel; vehicle manufacturers require reciprocal assurance about future fuel supplies. International experience, for example in Brazil and New Zealand, has demonstrated that significant

The most comprehensive analysis of CO_2 collection has ... suggested that collection and disposal of power-plant CO_2 emission would roughly double (at least) the cost of coal-fired electric power.

fuel switching in transportation can be accomplished but that it requires a concerted effort.

6.1.3 Capturing CO_2 Emissions

It is possible to burn hydrocarbon fuels and collect the CO_2 that is discharged. This sort of CO_2 collection is required for respired CO_2 in submarines and has been done at a variety of large chemical plants and fuel-burning facilities where there was a market for the CO_2. The basic technology is mature. In general, CO_2 is absorbed in a hydrous amine solution, and the working amine solution is regenerated by heating to drive off the CO_2. Carbon dioxide collection from the ambient atmosphere would require more energy than is made available during the generation of CO_2 in the first place, but the process is energetically feasible at an exhaust stream where the CO_2 concentration would be in the 10 to 13% range or higher. The process has been investigated in connection with such large-volume CO_2 users as enhanced oil recovery projects. It is also possible to collect CO_2 by refrigeration, and there has been some effort to develop gas-selective membranes that would separate the CO_2 from an exhaust stream. Although removal and recovery are possible, the problem remains what to do with the CO_2 once collected.

The most comprehensive analysis of CO_2 collection has been undertaken by Steinberg et al. (1984) who suggested that collection and disposal of power-plant CO_2 emission would roughly double (at least) the cost of coal-fired electric power. Disposal of the CO_2 was accomplished by injection below the oceanic thermocline, an idea first suggested by Marchetti (1975) in the mid-1970s. The CO_2 would be pumped into the ocean where it would mix into the high-density deep waters and be out of sight for a time scale on the order of the mixing time of the deep ocean (hundreds of years). The potential ecological impact has not been examined. Injection of some of the carbon into depleted oil and gas wells has been suggested, as has storage as solid CO_2 (dry ice) at the South Pole, injection into space, and a variety of other creative but impractical schemes. There is no metal cation in the Earth in sufficient quantities to permit storage of CO_2 as a solid carbonate and no marketable commodity that can be made from CO_2 in capacities required to be of significance.

Another limitation of this sort of CO_2 collection is that only about one third of global emissions are from large stationary sources where the necessary processing and collection equipment could even be contemplated. Marchetti (1975) has suggested that this fraction could be raised substantially by using, for example, large central facilities to convert hydrocarbon fuels into hydrogen fuels for small and/or mobile emitters. In this way, the bulk of carbon emissions would be concentrated at a smaller number of sites, although there would be an increase in the total amount of CO_2 generated.

Another option that has been examined in detail is combusting fossil fuels with pure oxygen and injecting the resulting 95% CO_2 flue gas stream directly into oil fields to enhance oil recovery (Wolsky and Brooks, 1987). Such an approach has the advantage that no gas separation is required and there are no other environmental emissions because the trace quantities of SO_2 and fuel-nitrogen based NO_x are injected along with the CO_2. The concept has been proven both

> **The most economically attractive way of collecting carbon from the ambient atmosphere (or from exhaust streams) is to use solar energy and to take advantage of biological photosynthesis ... A serious limitation is that the growth rate is highest for young vigorous trees, and with time the growth rate and CO_2 uptake will decline.**

in experimental boilers and in field experiments using existing utility coal-fired boilers. It is estimated that at oil prices of $30/barrel and above the process could be economically self sufficient in selected areas of the country which contain oil fields that could benefit from secondary recovery (Sparrow et al., 1988).

There has also been some discussion of processing fuels so that full fuel oxidation does not occur, and the carbon-bearing product is something other than CO_2. Steinberg (1989a) for example, discussed the possibility of processing coal so that the final products are carbon black and H_2O. This process extracts from coal only the energy available from hydrogen oxidation, but it yields a solid carbon product that should be easier to store than would gaseous or dissolved CO_2. Steinberg's estimate is that 24% of the energy value of coal would be available in such a scheme, and the fraction would be higher for liquid and gaseous fuels with higher hydrogen content.

The most economically attractive way of collecting carbon from the ambient atmosphere (or from exhaust streams) is to use solar energy and to take advantage of biological photosynthesis. Growing plants convert atmospheric CO_2 into solid, carbon-bearing compounds, and fast growing plants growing under optimum conditions can remove large quantities of carbon from the atmosphere. The primary question confronting large-scale plant uptake—generally perceived as an increase in global forest mass—is a matter of scale. How big a contribution could forests make in taking up fossil-fuel-released CO_2?

Other forms of biomass are possible, and both herbaceous crops and algae systems are being evaluated as sources of fossil-fuel substitutes, but if our interest is in storing carbon in standing biomass, then trees are the most attractive possibility. Forestry programs could involve either establishing new forest areas or increasing the productivity of existing forest areas. It has been widely cited that to take up all fossil-fuel emissions in fast-growing plantation forest (15 tons dry matter per hectare per year) would require new forest area equivalent to the area of Australia. Although this suggests that a full offset is not practical, it also suggests that a partial offset is feasible, and one U.S. power company has already initiated an offset program for a new coal-fired power plant. The trees are to be planted in a developing country.

A serious limitation is that the growth rate is highest for young vigorous trees, and with time the growth rate and CO_2 uptake will decline. As a consequence, it is necessary either (1) to envision reforestation as an interim measure that buys time while further understanding is achieved or energy systems are converted to nonfossil alternatives or (2) some way must be envisioned to harvest and store, without oxidation, the standing crop while a new crop is initiated. The merits of option 1 are both obvious and realistic, while the logical outgrowth of option 2 is that the harvested wood would be used to replace fossil fuels in what becomes a biomass fuels approach—although there will be a significant, long-term, net storage of carbon in the growing crop. Were wood to be the primary construction material, this would contribute to long-term storage of carbon and might encourage further forestation.

There are large areas in the U.S. and around the world that appear to be amenable to reforestation, although we do not yet have a good idea of what kinds of productivity might be achieved on the widely variable sites or how long that level of productivity could be sustained. Any large-scale forestry scheme would require the participation of a number of developing nations and raises many social and institutional as well as technical issues.

6.1.4 Preserving Global Forests

It has been estimated that 20 to 30% of current CO_2 emissions come not from energy systems but from a decrease in the mass of carbon in the biosphere (see Section 2.2.2.). Currently, the primary source is believed to be the clearing of tropical forests, but there remains uncertainty (e.g., see Pearman and Hyson, 1986; Pearman, 1989). Although planting additional forest land could offset some fossil-fuel-related CO_2 emissions, preserving existing forests would reduce net emissions. Current forest clearing is largely being done in developing tropical countries and is being driven by a multitude of forces ranging from colonization of new areas to the need for additional cropland and the expansion of infrastructure. Controlling these emissions will involve addressing the social and economic causes of forest clearing.

6.1.5 Controlling Population Growth

A continuing problem with reducing emissions from fossil-fuel systems is that any gains from improvements in end-use efficiencies, for example, can be quickly offset by increases in population. So long as population continues to increase, we can anticipate that there will be increasing pressure on fossil-fuel use and forest clearing and decreasing resilience to climate or social change. From 1950 to 1987, global total CO_2 emissions from fossil-fuel burning and cement manufacture increased by a factor of 3.45. During this same interval, CO_2 emissions per capita increased by a factor of 1.72. This suggests that, to a first approximation, over the last 37 years, CO_2 emissions have increased by a factor of 1.72 because of increases in personal energy consumption and standard of living and by a factor of 2.0 (3.45/1.72) because of increases in population. Technologies exist for safely controlling population growth, but their availability may be limited by economic factors and public acceptance.

To a first approximation, over the last 37 years, CO_2 emissions have increased by a factor of 1.72 because of increases in personal energy consumption and standard of living and by a factor of 2.0 because of increases in population.

6.1.6 Conservation/Changing the Structure of Demand

Most demand for energy is derived from demand for other goods and services whose delivery requires consumption of energy. All else being equal, reducing the demand for these goods and services, or substituting other goods and services whose production requires less energy, or improving the energy efficiency with which a given good or service is supplied, will reduce the demand for energy and its associated greenhouse gas emissions. The first of these approaches, efficiency, has already been discussed. Several strategies appear especially promising under the other approaches of reducing demand or substituting other goods and services.

- Recycling materials, recovering resources and reducing waste.
- Substituting low-emission materials for high-emission materials.
- Redesigning products and production processes to require less material.

All else being equal, reducing the demand for goods and services, or substituting other goods and services whose production requires less energy, or improving the energy efficiency with which a given good or service is supplied, will reduce the demand for energy and its associated greenhouse gas emissions.

- Land-use and business planning to reduce length of trips or number of trips requiring use of automobiles and aircraft.
- Designing communities and housing to reduce demand for space conditioning and water.
- Adjusting water prices to reduce highly consumptive uses and the associated energy for pumping.
- Reducing population growth to reduce future increase in demand for goods and services.
- Developing and substituting recreational activities that require less energy than some now popular activities.

In general, technologies exist to implement each of these strategies. However, low-emission substitutes do not exist for most uses of cement and for many applications of other energy-intensive materials, although use of fly-ash as an additive to portland cement does reduce emissions by 15 to 20%. Research and development would be needed to develop such substitutes. Research and development are also needed to develop good, moisture-resistant substitutes for CFC-foam insulation used in new buildings and appliances and allow substitution for these materials in new installations without increasing energy consumption. Each of these strategies would require the consumption of energy to implement, and additional information would be needed to determine the desirability of specific options.

There are significant political, institutional, and behavioral obstacles to implementing many of these strategies. With few exceptions, the U.S. political culture prefers to leave demand for goods, services, and activities up to individuals and businesses, and does not consider public policies (or tax laws) explicitly formulated to influence demand to be legitimate. Thus, strategies to influence demand may require justification on other grounds, although some of the strategies mentioned above (waste reduction, land-use planning) would respond to public concern over solid waste disposal, housing costs, and traffic congestion. However, the public does not yet appear to be concerned with other problems that would require reducing population growth or substituting leisure activities that require less energy as important parts of the solution, at least in the U.S.

Because there is little public-sector experience with demand modification, it is likely that institutions would take some time to develop the expertise needed to be effective; during this time, their effectiveness probably would be offset, at least in part, by private-sector firms that have a stake in the activities to be discouraged (high-emissions materials, recreational travel).

Given the complexity of individual decision-making, it is possible that a policy to modify demand for one activity might do so but lead simultaneously to other decisions that offset the original intent of the policy. For example, it is conceivable that people might find a community designed for energy efficiency to be restricting and thus seek additional stimulation by increasing their demand for recreational travel outside the community. Hence, before the policy is implemented, careful analysis is required to assess the likely net effects of the policy and its second-round or indirect effects. Monitoring of the policy's effects is required subsequently. It is also common for a policy to require different methods of implementation with different parts of the population. An energy tax or price increase might, for example, reduce demand for energy-intensive activities by some segments of the economy (e.g., the poor) but not the demand by other segments of the economy (e.g., the rich).

6.2 Change-Options with Nonfossil Sources of Energy

Over the long term, maintaining low levels of CO_2 emissions will depend on the availability of nonfossil energy sources that do not discharge CO_2 to the atmosphere.[2] It is generally conceded that there is, at present, no single energy technology with competitive price, acceptable social and environmental impact, and resources of sufficient magnitude to offer an acceptable alternative to fossil fuels. There are several alternatives, however, that, alone or in combination, have the potential to fill such a role. Nuclear fission, advanced biomass technologies, hydropower, and perhaps eventually nuclear fusion are obvious candidates. Although hydropower has limited expansion potential in the U.S. because the best sites have already been developed, its expansion in developing countries is potentially large. Photovoltic, solar-thermal, wind, and geothermal technologies in combination will likely play a significant role. With the exception of geothermal energy, these technologies produce no CO_2 during operation, their only contributions being embodied in plant capital (e.g., cement and construction), land-use changes (e.g., drowning forests behind hydroelectric dams), fuel supply (e.g., uranium mining), and the emissions that may be associated with necessary backup systems (often fossil-fuel powered) when these systems are not on line.

It is generally conceded that there is, at present, no single energy technology with competitive price, acceptable social and environmental impact, and resources of sufficient magnitude to offer an acceptable alternative to fossil fuels. There are several alternatives, however, that, alone or in combination, have the potential to fill such a role.

The size of the role that nuclear power plays in reducing greenhouse gas emissions will likely depend on how successfully society is able to grapple with the socioeconomic issues. Nuclear power plants emit no CO_2 during operation. Nuclear is a commercially available technology that already plays a large and expanding role in world energy supply. There are more than 400 commercial nuclear reactors worldwide and more than 100 in the U.S. producing approximately 20% of our electricity. However, there are presently no new orders for nuclear power plants in the U.S. Among the factors which have limited the growth of nuclear power are the unpredictabililty of construction costs, a deceleration in demand growth rate, and prudency litigation during the late 1970s and early 1980s; safety and waste storage concerns have also played a role. The reception of nuclear energy elsewhere in the world has been mixed. While France produces most of its electricity by nuclear generation, Sweden has decided by popular referendum to phase out its nuclear program. Studies indicate that many factors other than public perception of health and safety risks are relevant here, including issues of cost, institutional trust, distributive justice, and mechanisms of societal

[2] See also the companion multi-laboratory report on alternative energy technologies for a discussion considering more than aspects related to greenhouse gases.

> **The size of the role that nuclear power plays in reducing greenhouse gas emissions will likely depend on how successfully society is able to grapple with the socioeconomic issues.**

consent. In many ways these constraints are less tractable than technical constraints.

There are several advances in nuclear plant design on the horizon that hold the potential to have a significant impact on the U.S. nuclear industry. One involves an advancement of current light water reactors (LWR) generally called advanced LWR's. These reactors are similar to current reactors but with enhanced safety features and increased operating efficiencies. Another future nuclear option involves advanced or second generation reactor designs with passive safety characteristics such as the PIUS design or the high temperature gas reactor (HTGR). Another example is a breeder reactor under development known as the advanced liquid metal reactor (LMR). The advanced LMR differs from other reactor concepts in that it is an entire reactor system that treats the reactor, fuel cycle, and waste systems in a comprehensive manner. In doing so, problems of safety, waste disposal, proliferation, and costs are naturally addressed. The basis of the process is a new type of metal fuel that is submerged in a liquid sodium pool as opposed to a conventional loop cooling system. The design is such that even if all reactor cooling is cut off, the core naturally cools itself down. The metal fuel design enables the reactor to be coupled to on-site reprocessing, which recycles the plutonium and other long life radioactive by-products and leaves a waste that decays to that of natural uranium in a few hundred years as compared to several thousand years with conventional

> **Advanced biomass systems have the potential to supply large amounts of energy ... the possibility of high efficiency, sustainable, no-net-CO_2, electric power systems. Biomass systems offer the further advantage of being a potential source of liquid transportation fuels.**

reactor waste. On-site processing also prevents unauthorized fuel diversion and elimnates the need to transport radioactive material to or from the site during the reactor's lifetime. Costs are reduced because the inherently safe design and ease of fuel reprocessing provides for simplification and standardization of plant construction and operation.

Advanced biomass systems have the potential to supply large amounts of energy—Fulkerson (1989) suggests 14 quads of liquid fuels in the U.S.—for both developed and developing countries. Combined with increasing growth rates on experimental plantations, systems such as the whole tree burner of Ostlie (1988) and the aeroderivative gas turbines of Williams (1989) offer the possibility of high efficiency, sustainable, no-net-CO_2, electric power systems. Biomass systems offer the further advantage of being a potential source of liquid transportation fuels. Biomass is already a major source of energy globally but is often used with very low efficiency and in a manner that is resulting in net depletion of the resource. Advanced biomass fuels and improved photovoltaics offer additional advantages in low emission rates for other pollutants, plus domestically secure energy sources for many nations.

Photovoltaic energy sources are presently competitive with other energy sources for many remote applications both foreign and domestic, because its life-cycle cost is below that of competing electric power technologies. It is competitive with centralized power for

Photovoltaic energy sources are presently competitive with other energy sources for many remote applications both foreign and domestic, because its life-cycle cost is below that of competing electric power technologies.

remote applications because costs for grid extension are high.

It is anticipated by U.S. D.O.E. that advances in wind energy systems could soon lead to wind energy that would be competitive as a fuel displacer (no capacity credit) with oil and natural gas power generation as oil and gas prices increase. It may even compete with coal in many locations, particularly if the cost of coal includes environmental costs.

Technological refinements in solar thermal electric installations are expected to reduce cost by 30%. Solar thermal electric technology will benefit from applied research and development in the areas of reflective surfaces, stable absorbing surfaces, two-phase heat transfer related to durability issues, and heat engines. This research and development would emphasize durability and reliability which contribute to lower life-cycle cost.

U.S. geothermal electrical generation capacity will reach 3 GWe by the end of 1989, and geothermal sources contribute significant energy for heating applications. Three main technical limitations inhibit wider use of geothermal energy: uncertainty in reservoir characteristics; conversion system needs; and, the cost of drilling and completing a well.

The use of solar power satellites was considered about ten years ago, the recommendation being to wait ten years and to reconsider this source. If such energy could be transferred from space to the surface in a remote area (in order to limit environmental effects) and the energy is used to generate hydrogen

Advanced biomass fuels and improved photovoltaics offer additional advantages in low emission rates for other pollutants, plus domestically secure energy sources for many nations.

or is transmitted via high voltage lines, the potential exists over the long-term to provide large amounts of energy with little environmental cost; the economic costs of this approach, however, remain uncertain.

The development of nonfossil-fuel technologies, as part of a program to reduce the magnitude of global warming, will require understanding the advantages and specific obstacles faced by each technology (see companion multi-laboratory report). The full external costs (proxy costs associated with environmental damage, energy security, balance of trade, resource depletion, etc.) associated with each energy system (including fossil-fuel systems) must be determined to allow alternative systems to compete on an equitable basis. The concept of total fuel cycle cost will also allow comparisons across the spectrum of other options for dealing with potential climate changes.

6.3 Options for Tinkering with the Climate System

Because the Earth's climate and climate patterns are controlled by the incoming flux and internal redistribution of solar energy, it is theoretically possible to alter the climate and/or climate patterns by purposeful tinkering with the energy fluxes. For low and middle clouds, satellite data indicate that the cooling effect induced by cloud reflection of solar radiation exceeds the warming effect induced by cloud trapping of infrared radiation (Ramanathan et al., 1989). Climate model studies of volcanic aerosol effects and of the albedo feedback observed as a consequence

The full external costs (proxy costs associated with environmental damage, energy security, balance of trade, resource depletion, etc.) associated with each energy system (including fossil-fuel systems) must be determined to allow alternative systems to compete on an equitable basis.

of melting ice and snow suggest that significant effects could be engineered. According to Ramanathan (1988), an increase in planetary albedo of just 0.5% is sufficient to halve the greenhouse effect of a CO_2 doubling. Because only about half of the incoming solar radiation passes through the atmospheric gases and clouds to reach the Earth's surface, a purposeful albedo change would be most practically done as high as possible in the atmosphere. It has been suggested, for example, that because stratospheric aerosols reflect energy to space and cool the climate, injection of such aerosols could help limit climatic warming (e.g., Budyko, 1974; Broecker, 1985). Such an undertaking would, however, be massive. The injection of about 10^7 tonne of SO_2 by the El Chichon volcanic eruption in Mexico on April 4, 1982, has been estimated to be capable of causing a maximum surface cooling of about 0.3 to 0.4°C (Ramanathan et al., 1987). Aerosols in the lower atmosphere can also induce cooling, Wigley (1989) has suggested recently that SO_2-derived aerosols in the lower atmosphere of industrial areas may already be playing a cooling role in global climate.

Other possibilities that have been mentioned in this area include: large satellite systems that reflect away several percent of the incoming solar energy, and monolayers to retard evaporative losses from free-water surfaces. We could even tinker with sea level, to a point. It has been estimated that the large volume of water stored in freshwater reservoirs across the Earth has resulted in a millimeter-scale decrease in sea level. Spraying sea water onto the East Antarctic ice sheet could also help decrease sea level, but the energy cost might lead to warming that counteracts this effect.

To whatever extent we succeed in slowing the rate of change, we preserve our options and flexibility for future action.

The wisdom of this kind of purposeful climate intervention has been challenged repeatedly, and Broecker (1985) describes it as generally "taboo." The possibility of large interactions or unforeseen consequences is certainly great, but Broecker and others have wondered if the schemes might merit closer scrutiny, given the possibility there might be a low-cost scheme and that these interventions could generally be quickly halted (and their effects reversed) if they failed to function as anticipated.

6.4 Options for Coping with Climate Change

Coping with climate change could be consciously selected. It is also the default position if we decline to choose some other option; as such, it is an option that has already been selected to some degree. The CO_2 concentration in the atmosphere is 25% above the preindustrial value and is likely to increase further as we evaluate the effects and consider the options and urgency for other action. Any actions taken now to reduce emissions will slow the rate of change of atmospheric composition, slow the rate of climate change, and lengthen the time interval over which adaptive strategies would need to be implemented. To whatever extent we succeed in slowing the rate of change, we preserve our options and flexibility for future action.

A serious consequence of relying on coping strategies is the likelihood that the impact will be unequally distributed and the certainty that different countries and peoples are unequally equipped to adapt. In general, the poor are least able to adapt, and the rich must evaluate their responsibility to assist. Adaptations that may be quite feasible in the U.S. may be out of the question in a country like Bangladesh. For the poor, one of the best protections against change may be accelerated economic development, a process that would likely be a positive feedback to climate change by stimulating energy use.

Another serious consideration of coping strategies has to do with planning. In many economic transactions there is a long-term commitment, and the need to anticipate correctly is fueled by the high cost of error. Electric utilities, for example, must be able to anticipate the regional and temporal distribution of demand, the availability of hydropower, and the reliability of cooling water. Changes in the pattern of temperature and precipitation could have a major impact on each of these factors, and the utility would confront major costs if it either failed to anticipate change or prepared for a change that did not occur.

It is in this spirit that society confronts the whole issue of planning for climate change, of beginning to adopt energy policy, for example, to deal with anticipated changes but with a recognition of the cost of being wrong.

6.4.1 Dealing with Sea Level Rise

Changes in sea level are likely to occur very slowly, and many sectors of the advanced economies undoubtedly could respond by armoring the coast, relocating structures, importing water to replace resources lost to sea-water intrusion, etc., even though this might involve large costs if, for example, major items of long-term capital (power plants, hotels) were threatened. The same options are not available in developing countries with large populations in coastal areas or in island nations. In these two cases, adaptation would likely mean that other areas, other nations, would have to be prepared to accept what are coming to be called "environmental refugees." Other areas of concern involve impacts on coastal wetlands and reef structures.

6.4.2 Dealing with Temperature and Precipitation Changes

Technological responses to temperature and precipitation changes will include a major re-evaluation and perhaps restructuring of water management. Both physical structures and institutional arrangements to maintain or redistribute water for irrigation, water supply, hydropower, navigation, cooling, waste processing, fisheries, flood control, and recreation would be required. These might, in turn, feed back into CO_2 emissions by changing energy demand for irrigation, air conditioning, overland transport, etc. There could be a need to re-evaluate cropping patterns or crop varieties. In the field of agriculture, there could also be new opportunities if the direct effects of increased atmospheric CO_2 produced significant effects on productivity, drought tolerance, etc. Both within and among nations, there could be a significant reconfiguration of the distribution of goods and services. The demands on institutions could be severe.

6.4.3 Assisting Nature to Adapt

Natural systems are generally resilient to change, but there is concern that a human-inspired change might occur so rapidly as to tax the ability of systems to respond. As stated above, we do not know if coastal wetlands and coral reefs can respond rapidly enough to be preserved during an accelerated sea level rise to levels not experienced in more than 100,000 years or how we might assist if they cannot. For terrestrial ecosystems, we

Natural systems are generally resilient to change but there is concern that a human-inspired change might occur so rapidly as to tax the ability of systems to respond.

might be better able to assist adaptation or relocation, although the isolated nature and lack of continuity of some ecosystems may make it impossible for them to migrate or accommodate. We might have to respond to a changing frequency of wildfires, a decrease of snowfall in recreational areas, a change in the distribution of pests and pathogens.

6.5 Discussion

One of the fundamental questions in considering technological options is the relationship between CO_2 emissions and the change in atmospheric CO_2 concentration. In particular, authors like Firor (1988) and Harvey (1989) have suggested that there may be a CO_2 ration—a rate of fossil-fuel CO_2 emissions at which there will be no further increase in atmospheric CO_2. If there is such a ration, and at present estimates of its magnitude are quite uncertain (e.g., see Lashof and Tirpak, 1989), it is important to understand how large it is and what benefit there might be from reducing emissions to various levels.

Matters are further complicated because global climate change will be impacted by a variety of gases other than CO_2. As a result, control strategies for these other gases must be considered along with the CO_2 strategies, taking into account the very different relative contributions of each species to the direct climatic effects on infrared radiation and the indirect climatic effects through perturbing of atmospheric chemistry.

Trying to control emissions of greenhouse gases other than CO_2 will involve many different economic activities, many of them in the agricultural sector. Although most of these

Carbon dioxide ... is an essential and thermodynamically stable product of fossil-fuel combustion, and its generation can be completely avoided only by declining to burn fossil fuels or using only the hydrogen content in them.

activities are outside of the energy system, we should not forget that many of the control strategies will not be energy neutral. Adopting refrigerant working fluids to replace CFC-11 and CFC-12, for example, may adversely impact the efficiency of cooling systems. Production of nitrogen fertilizer consumes large quantities of energy, and if N_2O control strategies involve controlling the quantity or application technique for fertilizer nitrogen, there could be a significant energy feedback. Attempts to minimize methane leakage from energy systems or to capture methane, for example at coal mines or landfills, would likewise have energy-systems implications. In this context it is important to remember that emissions of methane, N_2O, the chlorofluorocarbons, and other trace species are fundamentally different from emissions of CO_2. For all of the other man-made greenhouse gases, emissions represent the undesired escape of a gas we should rather have retained or the inadvertent production and release of a gas that is incidental to the economic activity involved. Carbon dioxide from fossil fuels, on the other hand, is an essential and thermodynamically stable product of combustion, and its generation can be completely avoided only by declining to burn fossil fuels or using only the hydrogen content in them.

Technological options to deal with any of the greenhouse gases cannot be evaluated in isolation from the socioeconomic system of which they are a part. Costs and public acceptance barriers to technology penetration or behavioral change represent constraints

A key to evaluating technological responses to climate change is recognizing that it is a global problem that will require global solutions. There is no single technological fix, and no one nation can respond unilaterally to "solve" the problem.

that are just as binding as materials or thermodynamic efficiency constraints. In particular, public acceptance has been demonstrated to depend upon much more than understanding probabilities and magnitudes of unwanted consequences or economic benefits and costs. The behavioral sciences must be involved at an early stage in evaluations of technological options to avoid critical delays in implementation and unforeseen costs of conflict.

One implication of the imperative to understand technological options in their broader socioeconomic context is the linkage or tie-in strategy for responding to climate change. Some of the options for responding to climate issues will bring other benefits to society. For example, improvements in energy efficiency could improve the industrial competitiveness of the U.S. Tree planting to mitigate urban heat islands (thereby reducing electricity demand for cooling) may bring much appreciated aesthetic benefits. Movement away from dependence on hydrocarbon transport fuels and increased use of biomass for electrical generation could bring benefits in terms of U.S. energy security.

Similarly, initiatives pursued for reasons other than reducing greenhouse gas emissions may have beneficial effects on global warming. A prime example is the Montreal Protocol for the Protection of the Ozone Layer, which is designed to reduce CFC emissions on a global scale. However, it must be borne in mind that not all linkages or interactions between technologies are likely to be beneficial. Existing power-plant emission controls can increase CO_2 emissions per kilowatt-hour of electricity produced, and many potential technological options have negative implications for economic growth in developing nations.

A broad association with other goals should not be taken as a guarantee of efficiency and success. Such an association may increase the likelihood of unintended and undesirable effects as well. Care must be taken to distinguish (1) the advantages of recognizing positive externalities for climate that are generated by the pursuit of other goals from (2) the dangers of employing potential climate change as an argument to correct unrelated market failures. Whatever the merits of particular linkages between technologies, it makes sense to think strategically about the issue and not to isolate climate goals from other environmental and societal objectives.

A key to evaluating technological responses to climate change is recognizing that it is a global problem that will require global solutions. There is no single technological fix, and no one nation can respond unilaterally to "solve" the problem. Improvements in the efficiency of energy conversion and use, along with nonfossil energy systems and the protection and replanting of forests can decrease the rate of change. The challenges are technological, social, economic, and institutional. And, if the search for the most productive responses takes a least-cost approach that is cognizant of the many environmental and other linkages, it should acknowledge that everyone will not (indeed, need not) respond in the same way.

Chapter 7: Climate Change and the National Energy Strategy

Human activities over the past few centuries have begun to change atmospheric composition and climate. The atmosphere can no longer serve as the repository for the increasing amount of human-induced emissions of CO_2 and other chemically and radiatively active gases without leading to consequences that will modify climate, ecosystems, agriculture, shorelines, and many other aspects of our natural and created environment. It appears that many of these modifications will have adverse effects, especially if the rate of gaseous emissions increases and accelerates the rate of climate change. However, other of the modifications may be beneficial to specific groups or activities, making it more difficult coming to agreement on response strategies.

Certainty about the projected changes ranges from modest to very tentative, but the collective total of potential impacts could be substantial, generally irreversible, and quite widespread. The benefits of the activities generating these emissions are in many cases also substantial; we cannot suddenly stop activities such as the use of fossil fuels for energy generation and transportation and the production of cattle and paddy rice for food. There is also no single or easy technological fix that could suddenly stop the progression of the global climatic change that has been initiated. To maintain a continuing increase in the global standard of living will take a concerted effort, beginning with some readily available options, to even moderate the growth in emissions and begin to slow the rate of global change. This first step would also buy time for the longer-range and generally more expensive adjustments needed to reduce the potential consequences to at least within a level—as yet undefined—to which society and the environment can readily adapt without eventually leading to catastrophic changes, such as destruction of the polar ice caps.

This chapter summarizes the changes in atmospheric composition and climate that are, in turn, leading to the environmental impact of most concern. It also outlines the challenge posed if alternative energy technologies and approaches are to be helpful in addressing the greenhouse issue through development of the National Energy Strategy.

7.1 Potential Changes in Atmospheric Composition and Climate

7.1.1 Atmospheric Composition

That the atmospheric concentrations of carbon dioxide and other greenhouse gases are increasing is conclusively documented by an extensive observational record. Even though the emissions and budgets of these gases in the atmosphere-ocean-land-biosphere system are somewhat uncertain, there is nothing to indicate that the increasing concentration trends will do anything but continue upward, given present trends in human activities.

The concentration of carbon dioxide has increased about 25% during the period of industrial development and agricultural expansion beginning less than 200 years ago. Projections of accelerating development of the fossil-fuel resource, particularly coal, to raise the global standard of living suggest that the carbon dioxide concentration could reach double its preindustrial value by the middle-to-late decades of the next century.

The concentration of methane has more than doubled during the same period, and its rate of increase seems likely to continue as production of food and use of natural gas

Certainty about the projected changes ranges from modest to very tentative, but the collective total of potential impacts could be substantial, generally irreversible, and quite widespread.

continue upward. Although its absolute concentration is less than that of carbon dioxide, on a per molecule basis its effect on atmospheric radiation is about thirty times larger.

The concentration of chlorofluorocarbons has been increasing at relatively rapid rates. International action to reduce CFC emissions seems possible as substitute compounds are developed that will have much less effect on stratospheric ozone and somewhat less effect on the greenhouse effect.

The nitrous oxide concentration is also increasing, probably in large part because of the increasing use of fertilizers. Uncertainties are so large, however, that forecasts of concentrations are quite uncertain.

Together, these many emissions are also altering the natural concentrations of stratospheric and tropospheric ozone and of other compounds. These changes not only will enhance the greenhouse effect but also will increase the downward flux of ultraviolet radiation at the surface and affect the chemical cleansing capacity of the atmosphere.

7.1.2 Climate

Laboratory experiments and radiation model calculations demonstrate that increasing the concentration of these gases will intensify the trapping of upward infrared radiation and enhance the greenhouse effect, as occurred much earlier in the Earth's history and as now occurs, much more intensively, on Venus. The more difficult challenge is to estimate by how much this radiative change will alter the global and regional climate.

Although past climate changes can be used to enhance our understanding of the behavior of the climate system, the conditions under which these changes occurred and the uncertainty in details caused by the passage of time prevent their use as analogs. Climate models, tested against a variety of current and past climatic states, must therefore serve as the means for projecting the consequences of changes in atmospheric composition.

These climate models have both strengths and weaknesses. They are generally able to simulate the large-scale aspects of the global climate and the cycle of the seasons. However, the models do less well in reproducing climatic features on regional scales, particularly for precipitation and summertime temperature, in large part because their spatial resolution is quite poor and their representations of physical processes are still somewhat schematic.

The importance of environmental changes resulting from perturbed atmospheric composition has led to the use of these models to estimate the expected climatic changes, even though the models are still undergoing improvement to reduce recognized shortcomings. At their present stage of development, the models suggest that a rise in global temperatures to as much as a few degrees Celsius above current levels is likely during the next century if forecast emission trends are realized. Such a global change will significantly exceed the levels reached as a result of the decadal average natural variability to which society has become adapted. Associated with the greenhouse gas induced global change would be shifting precipitation and patterns of water availability, increasing evapotranspiration that would reduce summertime soil moisture, and rising sea level. Imposed on this rise in the mean temperature would be continued variability, making unusually warm (and possibly dry) conditions considerably more frequent.

Quantitative agreement among different models on estimates of these effects has not yet been satisfactorily achieved, perhaps because of shortcomings in the models, particularly concerning representation of clouds, chemistry, oceans, and ground hydrology. Thus, although the changes appear to be large, widespread, and persistent compared to the natural variations to which society has become adapted, the uncertainties remaining and the lack of information concerning changes in the frequency of extreme events limit and complicate the assessment of anticipated environmental effects.

7.2 Potential Environmental Impact

Many societal activities are closely tied to the weather and other high-frequency manifestations of the climate. A cost-benefit justification for altering emission patterns or for moving toward other strategies for modifying or adapting to climate change requires the identification of the major potential impact. Although little quantitative evaluation has yet been done, areas of consideration can be identified:

7.2.1 Food and Fiber Resources

Warming can lengthen the growing season, although major uncertainties concerning changes in precipitation and soil moisture limit our ability to predict the physiological effects of climate change on agriculture. Optimal locations for various crops may shift, increased evapotranspiration may aggravate water stresses and lead to summer drying of land (even desertification), water resources may be less abundant, and weeds and pests may be more competitive. The role of extreme events (e.g., hailstorms, which can devastate crops) cannot yet be evaluated. Also, there is not yet agreement on the direct role of increased CO_2 in enhancing photosynthesis and increasing water-use efficiency, given the current unknowns about interactions between effects of CO_2 and of climate and atmospheric pollutants, enhancement of weed competitiveness and herbivory, disruption of crop plant phenology and the potential for increases in leaf area to offset the gains from increased water-use efficiency.

Most assessments have assumed no adaptive response by the agricultural sector to climate change, and these assessments have often projected dire consequences. Although this may be the case in some of the low-technology countries of the developing world, in technologically and economically interconnected and progressive nations such as the U.S., the net results are probably dependent on the related rates of change of both climate and agricultural practice.

Relatively little attention has been paid to commercial forestry, where the potential for management response must be considered. Because of the 50 to 100 year interval between planting and harvest, forestry operations will have to consider environmental changes in their planning.

7.2.2 Natural Ecosystems

Although plants and animals can migrate in response to climate shifts, it is not known whether this migration can occur rapidly enough in unmanaged ecosystems to keep pace with the projected climate change and whether barriers to migration (e.g., urban or agricultural land use) can be overcome. Some investigators have predicted species extinctions and major shifts in ranges.

Forest growth simulations indicate that the response of unmanaged ecosystems to potential climate warming could be complicted by interactions with soil water availability and feedbacks with biogeochemical cycles. Also, climate-induced disturbances like drought (and resultant effects like wildfires) could accelerate ecosystem change faster than that from warming alone. Ecosystem simulations have not been conducted for other biomes (e.g., arid lands) and climate effects on ecological changes on terrestrial-aquatic linkages are largely unknown. Large-scale experiments and observations are needed to test and evaluate model predictions of ecosystem effects from proposed greenhouse warming.

There are long and complex causal chains linking climatic change with ultimate effects on fishery stocks. As a result, the relationship between climate variation and marine fisheries is not well known. The distribution of most fish species is expected to move poleward, but the magnitude of the shifts and implications for yields are not known. The role of abiotic factors in affecting the food supplies of immature fish is a major source of uncertainty.

7.2.3 Freshwater Resources

Shifting precipitation patterns (which are highly uncertain), increasing evaporation, higher elevation snowlines, and other factors seem likely to increase demand for water while reducing summertime water availability. Even though current water demands are already depleting groundwater in many areas, populations are expanding in many arid hydrologic basins (i.e., the Lower Colorado Basin), water allocation rules are frequently rigid, and wastes are reducing water quality. Climate-induced changes could further stress water resources and lead to more frequent and severe water-shortage episodes. Given the importance of freshwater to many other resource sectors (e.g., agriculture, industry, and municipalities), freshwater resources could have the most significant environmental impact in the U.S.

7.2.4 Rising Sea Level

Melting of mountain glaciers and thermal expansion of sea level are projected to raise sea level by up to one meter or possibly more by 2100 (compared with about 0.1 to 0.2 m over the past century) if current emission projections and models are correct. (As with temperature change, this projected range is not certain by about a factor of 2.) Although highly uncertain, there is the potential to initiate the deterioration of the Greenland and West Antarctic ice sheets. These ice sheets store the equivalent of 5 to 10 m of sea level, and the impacts of their decay, even though it is likely that it would take several centuries, would be profound. Coastal wetlands, aquifers, and structures would be threatened by the rising sea level, and these effects would be amplified by storm surges; warming-related increases in the frequency and intensity of tropical storms could further exacerbate the problem, although the nature of such changes cannot yet be quantified. Our understanding of coastal sediment transport is inadequate to predict patterns of coastal erosion and deposition, making it difficult to predict in detail how coastal regions will be impacted. The rate of sea level rise is as important as the eventual amplitude of the rise because (1) wetlands can adapt (by accreting vertically) if the sea level rise is not too rapid and (2) the lifetime (or replacement interval) of coastal structures would vary depending on the net rate of rise in sea level.

7.2.5 Human Health

Warmer summer temperatures can lead to increased heat stress. Warmer winter temperatures may reduce cold stress but may favor some insect-borne diseases. Much of the both hot- and cold-weather health stress is attributable to the onset of extreme episodes rather than to longer-term average conditions, but the variability of future climate is very uncertain. It seems possible that cold-related diseases such as influenza would decline with warmer weather while the incidence of hot-weather diseases could increase.

The net effect for most of the U.S. would probably be modest, but health effects could be of significant concern in semi-tropical and tropical portions of the U.S. and other low-latitude regions.

7.2.6 Variability and Extreme Events

Although models are not yet capable of projecting changes in variability of the frequency of extreme events (e.g., storms, floods, and droughts), there are indications that marginal and dry periods could be made drier and that hurricanes (which form preferentially over waters warmer than about 27.5°C) could be more frequent as ocean temperatures rise; conversely, extratropical storms (with their beneficial rains) could become less frequent as a result of a decreased thermal gradient between the poles and tropics. Such changes in variability could pose extreme impact on societal activities, many of which are reasonably well adapted to the present variability pattern.

7.3 Potential Societal and Technological Developments

Our present inability to accurately predict the future impact of climate change is owing in part to our uncertainty concerning future climate but also to the many uncertainties concerning the role of climate relative to other environmental factors (e.g., pollution), the linkages that connect regional systems of resources, and the future course of development of society and technology. It is important, while focusing on effects in the U.S., to bear in mind that this nation could also be affected by international developments caused by climate change. Given the economic, demographic, and commodity (e.g., agriculture, timber, and fossil-fuel) connectedness of nations, effects on individual nations cannot be realistically studied in isolation. Even though developing nations might be most vulnerable to the first-order effects of climate change, all nations may experience consequences.

Given the economic, demographic, and commodity connectedness of nations, effects on individual nations cannot be realistically studied in isolation.

The accepted linkage between fossil-fuel burning and the increasing concentration of atmospheric CO_2 focuses attention on the global energy supply system. It also establishes a primary role for the DOE. Innovative advances to improve the efficiency of energy conversion and use can slow the rate of increase of atmospheric CO_2 and allow more time to understand the consequences of increasing CO_2 and the necessity for more rigorous steps. Safe nuclear power, photovoltaics, advanced biomass systems, and other evolving technologies may provide a way to supply useful energy without adding greenhouse gases to the atmosphere.

With the exception of geothermal energy, the availability of all renewable energy sources depends in part upon climatic processes whose intensity and geographic distribution is now subject to changes in global climate. Changes in precipitation regimes could reduce hydroelectric resources in some regions where they are abundant and make them more available in others. Increases in cloud cover could limit the yield of photovoltaic or solar thermal technologies: decreases in cloud cover could make these technologies attractive in other regions. Changes in temperature, precipitation, and concentrations of CO_2 in the atmosphere can increase or decrease yields of biomass energy cropping systems, or require changes in crops to maintain comparable yields per hectare. Wind patterns and the yields of wind farms are also subject to change.

Investments in energy systems are long-lived, and the potential sensitivity of these resources to climate change adds to the uncertainty that all energy system planning must confront. Particularly worrisome is the possibility that human-induced climate change could reduce the ability of these resources to displace the energy sources that help to cause the change. This possibility has received surprisingly little attention.

All of these aspects require additional research to achieve technical, financial, and social goals, including a better understanding of the full environmental impact of advanced energy systems and of the full range of tradeoffs that might be made in selecting one system over another.

7.4 Climate Change and the National Energy Strategy

Scientific understanding of the potential climatic consequences is certain enough to indicate that warming of up to a few degrees is likely over the next hundred years if projected emissions trends are fulfilled. Such a warming would lead to conditions warmer than those experienced over the last hundred thousand years. Changes of this magnitude are generally not unprecedented on a short-term and local basis. However, the persistence of changes of climate of this magnitude, or changes in the frequency of warm (and often dry) conditions, may create unprecedented, but still quite uncertain, environmental consequences.

To be able to better estimate potential changes in climate and to evaluate the impact of such changes, we must learn more. A significant part, although not all, of our uncertainties can be reduced substantially by a concerted research effort. Development of the necessary knowledge base will not come easily—it will require substantial and sustained research efforts over many years, but it must be undertaken so that any actions taken now can be evaluated and adjusted later.

Many of the scientific aspects of the needed research are suggested in the recent plan developed by the Committee on Earth Sciences (CES, 1989) and in the concluding sections of Chapters 2 to 6 of this report. The Department of Energy, working through its Office of Energy Research, has already made many important contributions, and, working in cooperation with other government agencies, has responsibilities and capabilities that should provide the basis for further contributions to advancing scientific and societal understanding.

Although the situation is complex and somewhat uncertain, letting the uncertainties immobilize us may create at least as large an impact as moving ahead in a measured way with the extensive knowledge that is available. Secretary of Energy Admiral James Watkins has enunciated six principles to apply in developing the NES:

1. Take aggressive action on those issues on which scientific consensus exists.
2. Assess the state of the *science* on issues where *no* scientific consensus exists, and identify areas for further inquiry.
3. Where scientific uncertainty exists, move forward with those measures that make sense on other grounds, e.g., efficiency and reducing CFCs.
4. Consider the costs and benefits of any response measures suggested.
5. Link responses to scientific and technical information.
6. Determine how to evaluate and share technological responses with developing countries.

In testimony before the Senate Committee on Energy and Natural Resources, Admiral Watkins (1989) went on to say:

> It is entirely proper that the Department of Energy assume a leading role in

reducing the scientific uncertainty, and in defining the initiatives that can be undertaken—publicly and privately—in response to this threat.

The identification of aproaches and uncertainties in this report has been intended to suggest appropriate research to pursue.

Consideration of the greenhouse gas issue in development of national and international energy policy through the NES and through U.S. participation in the Intergovernmental Panel on Climate Change, will provide the potential for balancing the benefits and adverse impact from various courses of action. Although exceedingly complex because of scientific uncertainties, differing international perspectives on nature and technology, and the potential that changes in some regions and of some types may be beneficial, such consideration should be able to help optimize national and international choices. Economically attractive actions to slow the rate of emissions could provide time for the development of new technologies and make easier the adaptation to change. Developing a greater variety of economically and environmentally attractive energy systems is important in order to provide a broader set of options for future planning. Better understanding of potential changes would allow development of a greater resilience to changes.

Choosing to do nothing is just as much a choice as choosing to do something. This report has attempted to provide the technical basis upon which effective policies can be developed for consideration.

REFERENCES

Chapter 1

Bolin, B., B. R. Döös, J. Jaeger, and R. A. Warrick (eds.). 1986. *Greenhouse Effect, Climatic Change, and Ecosystems,* SCOPE 29, John Wiley and Sons, Chichester, U.K.

Budyko, M. I., A. B. Ronov, and A. L. Yanshin. 1987. *History of the Earth's Atmosphere,* Springer-Verlag, Berlin, 139 pp.

Gates, W. L. 1979. The Physical Basis of Climate, in *Proceedings of the World Climate Conference* (WMO No. 537), World Meteorological Organization, Geneva, Switzerland, 112–131.

Lashof, D. A. 1989. The Dynamic Greenhouse: Feedback Processes that May Influence Future Concentrations of Anthropogenic Trace Gases and Climatic Changes, *Climatic Change, 14,* 213–242.

MacCracken, M. C. and F. M. Luther (eds.). 1985a. *Projecting the Climatic Effects of Increasing Carbon Dioxide,* report DOE/ER-0237, U.S. Department of Energy, Washington, D.C., 381 pp.

MacCracken, M. C. and F. M. Luther (eds.). 1985b. *Projecting the Climatic Effects of Increasing Carbon Dioxide,* report DOE/ER-0235, U.S. Department of Energy, Washington, D.C.

NRC (National Research Council). 1983. *Changing Climate,* report of the Carbon Dioxide Assessment Committee, National Academy Press.

Smith, J. B. and D. A. Tirpak (eds.). 1988. *The Potential Effects of Global Climate Change on the United States, Draft Report to Congress,* U.S. Environmental Protection Agency, Washington, D.C.

Strain, B. R. and J. D. Cure (eds.). 1985. *Direct Effects of Increasing Carbon Dioxide on Vegetation,* report DOE/ER-0238, U.S. Department of Energy, Washington, D.C.

Trabalka, J. (ed.). 1985. *Atmospheric Carbon Dioxide and the Global Carbon Cycle,* report DOE/ER-0239, U.S. Department of Energy, Washington, D.C.

Chapter 2

CMA (Chemical Manufacturers Association). 1987. *Production, Sales, and Calculated Release of CFC-11 and CFC-12 through 1986.* Washington, D.C.

CMA (Chemical Manufacturers Association). 1988. *Production, Sales, and Calculated Release of CFC-11 and CFC-12 through 1987.* Washington, D.C.

Detwiler, R. P. and C. A. S. Hall. 1988. Tropical Forests and the Global Carbon Cycle, *Science, 239:* 42–74.

Dupont. 1987. Fluorocarbon/Ozone Update, Wilmington, DE.

Edmonds, J. and J. Reilly. 1985. Future Global Energy and Carbon Dioxide Emissions, *Atmospheric Carbon Dioxide and the Global Carbon Cycle,* report DOE/ER-0239, U.S. Department of Energy, Washington, D.C., pp. 215–246.

Edmonds, J. A., J. M. Reilly, R. H. Gardner, and A. Brenkert. 1986. *Uncertainty in Future Global Energy Use and Fossil Fuel CO_2 Emissions. 1975 to 2075.* report DOE/NBB-0081, U.S. Department of Energy, Washington, D.C. (available from NTIS, Springfield, Virginia).

Gifford, R. M. 1989. Direct Effects of Higher Carbon Dioxide Concentrations on Vegetation, in *Greenhouse Planning for Climate Change,* G. I. Pearman (ed.), CSIRO Publications, East Melbourne, Victoria, Australia.

Goldemberg, J., T. B. Johansson, A. K. N. Reddy, and R. H. Williams. 1987. *Energy for a Sustainable World,* Wiley-Easton, New Delhi, India.

Hao, W. M., S. C. Wofsy, M. B. McElroy, J. M. Beer, and M. A. Toqan. 1987. Sources of Atmospheric Nitrous Oxide From Combustion, *J. Geophys. Res. 92:* 3098–3104.

Houghton, R. A., J. E. Hobbie, J. M. Melillo, B. Moore, B. J. Peterson, G. R. Shaver, and G. M. Woodwell. 1983. Changes in the Carbon Content of Terrestrial Biota and Soils Between 1860 and 1980: A Net Release of CO_2 to the Atmosphere. *Ecological Monographs 53:* 235–262.

Houghton, R. A., R. D. Boone, J. R. Fruci, J. E. Hobbie, J. M. Melillo, C. A. Palm, B. J. Peterson, G. R. Shaver, G. M. Woodwell, B. Moore, D. L. Skole, and N. Myers. 1987. The Flux of Carbon from Terrestrial Ecosystems to the Atmosphere in 1980 due to Changes in Land Uses: Geographic Distribution of the Global Flux, *Tellus 39B:* 122-129.

Khalil, M. A. K. and R. A. Rasmussen. 1989a. The Role of Methylchloroform in the Global Chlorine Budget, Air and Waste Management Association paper 89-54.

Khalil, M. A. K. and R. A. Rasmussen. 1989b. Constraints Imposed by the Ice Core Data on the Budgets of Nitrous Oxide and Methane, Oregon Graduate Center, preprint.

Khalil, M. A. K. and R. A. Rasmussen. 1988. Nitrous Oxide: Trends and Global Mass Balance Over the Last 3000 Years, *Annals Glaciology 10:* 73.

Linak, W. P., J. A. McSorley, R. E. Hall, J. V. Ryan, R. K. Srivastava, J. O. L. Wendt, and J. B. Mereb. 1989. N_2O Emissions from Fossil Fuel Combustion, Air and Waste Management Association paper 89-4.6.

Mintzer, I.M. 1987. *A Matter of Degrees: The Potential for Controlling the Greenhouse Effects,* World Resources Institute, report #5.

Muzio, L. J. and J. C. Kramlich. 1988. An Artifact in the Measurement of N_2O from Combustion Sources, *Geophys. Res. Lett. 15:* 1369–1372.

Nordhaus, W. and G. Yohe. 1983. Future Carbon Dioxide Emisisons from Fossil Fuels, in *Changing Climate,* National Academy Press, Washington, D.C., pp. 87-153.

Rotman, J., H. de Boois and J. Swart. 1989. *An Integrated Model for the Assessment of the Greenhouse Effect: The Dutch Approach,* National Institute of Public Health and Enviornmental Protection, Rijksinstituut voor Volksgezondheid en Milieuhygiaene, Postbus 1, 3720 BA, Bilthoven, The Netherlands.

U.S. Environmental Protection Agency (EPA). 1989. *Policy Options for Stabilizing Global Climate,* D. A. Lashof and D. A. Tirpak (eds.), draft report to Congress, Washington, D.C.

Wahlen, M., N. Tanaka, R. Henry, B. Deck, J. Zeglen J. S. Vogel, J. Southon, A. Shemesh, R. Fairbanks, and W. Broecker. 1989. Carbon-14 in Methane Sources and in Atmospheric Methane: The Contribution from Fossil Carbon. *Science 245:* 286–290.

Watson, R. T. and Ozone Trends Panel, M. J. Prather and Ad Hoc Theory Panel, and M. J. Kurylo and NASA Panel for Data Evaluation. 1988. *Present State of Knowledge of the Upper Atmosphere 1988: An Assessment Report,* NASA Reference Publication 1208, Washington, D.C.

World Meteorological Organization, 1985. *Atmospheric Ozone: 1985,* Global Ozone Research and Monitoring Project, Report No. 16, Geneva.

Wuebbles, D. J. and J. Edmonds. 1988. *A Primer on Greenhouse Gases,* U. S. Department of Energy, Carbon Dioxide Research Division, report DOE/NBB0083.

Wuebbles, D. J., K. E. Grant, P. S. Connell and J. E. Penner. 1989. The Role of Atmospheric Chemistry in Climate Change, *Journal of the Air Pollution Control Association 39:* 22.

Chapter 3

AFEAS (Alternate Fluorocarbon Environmental Acceptability Study). 1989. United Nations Environment Programme, in press.

Bates, T. S., J. D. Cline, R. H. Gammon and S. R. Kelly-Hansen. 1987. Regional and Seasonal Variations in the Flux of Oceanic Dimethylsulfide to the Atmosphere, *Journal of Geophysical Research 92:* 2930–2938.

Blake, D. R. and F. S. Rowland. 1988. Continuing Worldwide Increase in Tropospheric Methane, 1978 to 1987, *Science 239:* 1129–1131.

Charlson, R. ., J. E. Lovelock, Mo. O. Andreae, and S. G. Warren. 1987. Oceanic Phytoplankton, Atmospheric Sulfur, Cloud Albedo, and Climate, *Nature 326:* 655–661.

Cicerone, R. J. 1988. How Has the Atmospheric Concentration of CO Changed?, in *The Changing Atmosphere,* F. S. Rowland and I. S. A. Isaksen (eds.) John Wiley and Sons Ltd.

Cicerone, R. J. and R. S. Oremland. 1988. Biogeochemical Aspects of Atmospheric Methane, *Global Biogeochem. Cycles 2:* 299–327.

Connell, P. S. and D. J. Wuebbles. 1989. Evaluating CFC Alternatives from the Atmospheric Viewpoint, Air and Waste Management Association paper 89–57, 1987, also Lawrence Livermore National Laboratory report UCRL-99927 Rev. 1.

DeLuisi, J. J., D. U. Longenecker, C. L. Mateer, and D. J. Wuebbles. 1989. An Analysis of Northern Middle-Latitude Umkehr Measurements Corrected for Stratospheric Aerosols for 1979–1986, *J. Geophys. Res.,* 9837–9846.

Edmonds, J. A. and J. Reilly. 1983. A Long-term Global Energy-economic Model of Carbon Dioxide Release from Fossil Fuel Use. *Energy Economics 5:* 74–88.

Edmonds, J. A., J. M. Reilly, R. H. Gardner, and A. Brenkert. 1986. *Uncertainty in Future Global Energy Use and Fossil Fuel CO_2 Emissions. 1975 to 2075.* report DOE/NBB-0081, U.S. Department of Energy, Washington, D.C. (available from NTIS Springfield, Virginia).

Emanuel, W. R., G. G. Killough, W. M. Post, and H. H. Shugart. 1984. Modeling Terrestrial Ecosystems in the Global Carbon Cycle with Shifts in Carbon Storage Capacity by Land-Use Change, *Ecolocy 65:* 970–983.

Emanuel, W. R., G. G. Killough, W. M. Post, H. H. Shugart, and M. P. Stevenson. 198. *Computer Implementation of a Globally Averaged Model of the World Carbon Cycle,* TR010, Carbon Dioxide Research Division, U.S. Department of Energy, Washington, D.C.

Enting, I. G. and J. V. Mansbridge. 1987. The Incompatibility of Ice-core CO_2 Data with Reconstructions of Biotic CO_2 Sources. *Tellus 39B:* 318–325.

Friedli, H., H. Lötscher, H. Oeschger, U. Siegenthaler, and B. Stauffer. 1986. Ice Core Record of $^{13}C/^{14}C$ Ratio of Atmospheric Carbon Dioxide in the Past Two Centuries. *Nature 324:* 237–238.

Hansen, J., I. Fung, A. Lacis, D. Rind, S. Lebedeff, R. Ruedy, G. Russell, and P. Stone. 1988. Global Climate Changes as Forecast by Goddard Institute for Space Studies Three-dimensional Model, *J. Geophys. Res. 93:* 9341.

Houghton, R. A., J. E. Hobbie, J. M. Melillo, B. Moore, B. J. Peterson, G. R. Shaver, and G. M. Woodwell. 1983. Changes in the Carbon Content of Terrestrial Biota and Soils Between 1860 and 1980: A Net Release of CO_2 to the Atmosphere, *Ecological Monographs* 53:235–262.

Isaksen, I.S.A. and O. Hov. 1987. Calculations of Trends in the Tropospheric Concentrations of O_3, OH, CO, CH_4, and NO, *Tellus 398,* 271–285.

Johnston, H., D. Kinnison and D. Wuebbles. 1989. Nitrogen Oxides from High Altitude Aircraft: An Update of Potential Effects on Ozone, *Journal of Geophysical Research* (accepted), also Lawrence Livermore National Laboratory report UCRL-100714, rev. 1.

Keeling, C. D. 1986. Atmospheric CO_2 Concentrations—Mauna Loa Observatory, Hawaii 1958–1986, NDP-001/R1, Carbon Dioxide Information Center, Oak Ridge National Laboratory, Oak Ridge, TN.

Keeling, C. D., R. B. Bacastow, A. E. Bainbridge, C. A. Ekdahl, P. R. Guenther, L. S. Waterman, and J. F. Chin. 1976. Atmospheric Carbon Dioxide Variations at Mauna Loa Observatory, Hawaii, *Tellus 28:* 538–551.

Khalil, M. A. K. and R. A. Rasmussen. 1987. Atmospheric Methane: Trends Over the Last 10,000 Years, *Atmos. Env. 21:* 2445–2452.

Khalil, M. A. K. and R. A. Rasmussen. 1988a. Carbon Monoxide in the Earth's Atmosphere: Indications of a Global Increase, *Nature 332,* 242–245.

Khalil, M. A. K. and R. A. Rasmussen. 1988b. Nitrous Oxide: Trends and Global Mass Balance Over the Last 3000 Years, *Annals Glaciology 10:* 73–79.

Killough, G. G. and W. R. Emanuel. 1981. A Comparison of Several Models of Carbon Turnover in the Ocean with Respect to Their Distributions of Transit Time and Age and Response to Atmospheric CO_2 and ^{14}C, *Tellus 33:* 274–290.

Komhyr, W. D., R. H. Gammon, T. B. Harris, L. S. Waterman, T. J. Conway, W. R. Taylor, and K. W. Thoning. 1985. Global Atmospheric CO_2 Distribution and Variations from 1968–1982 NOAA/GMCC CO_2 Flask Sample Data, *Journal of Geophysical Research 90,* 5567–5596.

Lacis, A., J. Hansen, P. Lee, T. Mitchell, and S. Lebedeff. 1981. Greenhouse Effect of Trace Gases, 1970–1980, *Geophys. Res. Lett. 8:* 1035–1038.

Lacis, A. A., D. J. Wuebbles, and J. A. Logan. 1990. Radiative Forcing of Global Climate by Vertical Distribution Change of Atmospheric Ozone, *J. Geophys. Res.,* in press.

Lashof, D. A. 1989. The Dynamic Greenhouse: Feedback Processes that May Influence Future Concentrations of Anthropogenic Trace Gases and Climatic Changes, *Climatic Change, 14,* 213–242.

Logan, J.A. 1985. Tropospheric Ozone: Seasonal Behavior, Trends, and Anthropogenic Influence, *J. Geophys. Res. 90:* 10463–10482.

MacCracken, M.C. and F.M. Luther (eds.). 1985. Detecting the Climatic Effects of Increasing Carbon Dioxide, report DOE/ER-0235, U.S. Department of Energy, Washington, D.C.

Marland G., T. A. Boden, R. C. Griffin, S. F. Huang, P. Kanciruk, and T. R. Nelson. 1989. *Estimates of CO_2 Emissions from Fossil Fuel Burning and Cement Manufacturing, Based on the United Nations Energy Statistics and the U.S. Bureau of Mines Cement Manufacturing Data.* ORNL/CDIAC-25. Oak Ridge National Laboratory, Oak Ridge, Tennessee.

Mintzer, I.M. 1987. *A Matter of Degrees: The Potential for Controlling the Greenhouse Effects,* World Resources Institute, report #5.

Mooney, H. A., P. M. Vitousek, and P. A. Matson. 1987. Exchange of Materials Between Terrestrial Ecosystems and the Atmosphere, *Bio-Science, 238,* 926–932.

National Research Council. 1988. *Toward an Understanding of Global Change,* National Academy Press, Washington, D.C.

Neftel, A., E. Moor, H. Oeschger, and B. Stauffer. 1985. Evidence from Polar Ice Cores for the Increase in Atmospheric CO_2 in the Past Two Centuries, *Nature 315:* 45–47.

Oeschger, H., U. Siegenthaler, and A. Gugelman. 1975. A Box Diffusion Model to Study the Carbon Dioxide Exchange in Nature, *Tellus 27:* 168–192.

Pearman, G.I. (ed.). 1989. *Greenhouse: Planning for Climate Change,* CSIRO Publications, East Melbourne, Victoria, Australia.

Pearman, G.I., D. Etheridge, F. DeSilva, and P.J. Fraser. 1986. Evidence of Changing Concentrations of Atmospheric CO_2, N_2O, and CH_4 from Air Bubbles in Antarctic Ice, *Nature 320:* 248–250.

Penner, J. 1989. Cloud Albedo, Greenhouse Effect, Atmospheric Chemistry, and Climate Change. Air and Waste Management Society paper. Also Lawrence Livermore National Laboratory report UCRL-99928 (submitted to *Journal of Air Pollution Control Association).*

Prinn, R.G., D. Cunnold, R.A. Rasmussen, P. Simmonds, F. Alyea, A.J. Crawford, P. Fraser, and R. Rosen. 1987. Atmospheric Trends in Methyl Chloroform and the Global Average for the Hydroxyl Radical, *Science 238:* 945–950.

Prospero, J. M., R. J. Charlson, V. Mohnen, R. Jaenicke, A. C. Delany, J. Moyers, W. Zoller and K. Rahn. 1983. The Atmospheric Aerosol System: An Overview, *Reviews of Geophysics 21:* 1607–1629.

Ramanathan, V., L. Callis, R. Cess, J. Hansen, I. Isaksen, W. Kuhn, A. Lacis, F. Luther, J. Mahlman, R. Reck, and M. Schlesinger. 1987. Climate-chemical Interactions and Effects of Changing Atmospheric Trace Gases, *Rev. Geophys. 25:* 1441–1482.

Ramanathan, V., R. J. Cicerone, H. B. Singh, and J. T. Kiehl. 1985. Trace Gas Trends and Their Potential Role in Climate Change, *J. Geophys. Res. 90:* 5547–5566.

Raynaud, D., J. Chappellax, J. M. Barnola, Y. S. Korotkevich, and C. Lorius. 1988. Climatic and CH_4 Cycle Implications of Glacial-interglacial CH_4 Change in the Vostok Ice Core, *Nature, 333:* 655–659.

Revelle, R. 1983. Methane Hydrates in Continental Slope Sediments and Increasing Carbon Dioxide, *Changing Climate,* National Research Concil, Washington, D.C., pp. 252–261.

Solomon, S. 1988. The Mystery of the Antarctic Ozone "Hole," *Reviews of Geophysics 26:* 131–148.

Tiao, G. C., G. C. Reinsel, J. H. Pedrick, G. M. Allenby, C. L. Mateer, A. J. MIller, and J. J. DeLuisi. 1986. A Statistical Trend Analysis of Ozonesonde Data, *J. Geophys. Res. 91:* 13121–13136.

Thompson, A. M., R. W. Stewart, M. A. Owen, and J. A. Jervehe. 1989a. Sensitivity of Tropospheric Oxidants to Global Chemical and Climate Change, *Atmos. Environ. 23:* 519–532.

Thompson, A.M., M.A. Huntley, and R.W. Steward. 1989b. Perturbations to Tropospheric Oxidants, 1985–2035: Calculations of Ozone and OH in Cemially Coherent Regions, submitted to *J. Geophys. Rev.,* 1989.

Trabalka, J. (ed.). 1985. *Atmospheric Carbon Dioxide and the Global Carbon Cycle,* report DOE/ER-0239, U.S. Department of Energy, Washington, D.C.

Trabalka, J. R., J. A. Edmonds, J. M. Reilly, R. H. Gardner, and D. E. Reichle. 1986. Atmospheric CO_2 Projections with Globally Averaged Carbon Cycle Models. pp. 534–560. In J. R. Trabalka and D. E. Reichle (eds.) *The Changing Carbon Cycle: A Global Analysis.* Springer-Verlag, New York.

U.S. Environmental Protection Agency (EPA). 1989. *Policy Options for Stabilizing Global Climate,* D. A. Lashof and D. A. Tirpak (eds.), draft report to Congress, Washington, D.C.

Wang, W.-C., D. J. Wuebbles, W. M. Washington, R. G. Isaacs, and G. Molnar. 1986. Trace Gases and Other Potential Perturbations to Global Climate, *Rev. Geophys. 24:* 110–140.

Watson, R. T., and Ozone Trends Panel, M. J. Prather, and Ad Hoc Theory Panel, and M. J. Kurylo, and NASA Panel for Data Evaluation. 1988. *Present State of Knowledge of the Upper Atmosphere 1988: An Assessment Report,* NASA Reference Publication 1208.

World Meteorological Organization. 1985. *Atmospheric Ozone: 1985,* Global Ozone Research and Monitoring Project, Report No. 16.

World Meteorological Organization. 1989. *Assessment of Stratospheric Ozone 1989,* report in progress.

Wuebbles, D. J. 1989. Beyond CO_2—The Other Greenhouse Gases, Lawrence Livermore National Laboratory report UCRL-99883; also published by the Air and Waste Management Association, paper 89-119.4.

Wuebbles, D. J., K. E. Grant, P. S. Connell, and J. E. Penner. 1989. The Role of Atmospheric Chemistry in Climate Change, *Journal of the Air Pollution Control Association 39:* 22–28.

Wuebbles, D.J. and D.E. Kinnison. 1988. A Two-dimensional Model Study of Past Trends in Global Ozone, *Proceedings Quadrennial Ozone Symposium,* University of Göttingen, West Germany.

Wuebbles, D.J., M.C. MacCracken, and F.M. Luther. 1984. *A Proposed Reference Set of Scenarios for Radiatively Active Atmospheric Constituents,* technical report DOE/NBB-0066, U.S. Department of Energy Carbon Dioxide Research Division, Washington, D.C.

Chapter 4

Arrhenius, S. 1896. On the Influence of Carbonic Acid in the Air Upon the Temperature of the Ground, *Phil. Mag. 41:* 237-276.

Arrhenius, S. 1908. *Worlds in the Making,* Harper, New York.

Barnett, T. P. and M. E. Schlesinger. 1987. Detecting Changes in Global Climate Induced by Greenhouse Gases. *J. of Geophysical Research 92,* 14772–14780.

Barnola, J. M., D. Raynaud, Y. S. Korotkevich and C. Lorius. 1987. Vostok Ice Core Provides 160,000-year Record of Atmospheric CO_2, *Nature 239:* 408–414.

Barth, M. C. and J. G. Titus (eds.). 1984. *Greenhouse Effect and Sea Level Rise: A Challenge for this Generation,* Van Nostrand Reinhold, New York.

Budyko, M. I., A. B. Ronov, and A. L. Yanshin. 1985. *History of the Earth's Atmosphere,* Gidrometeoizdat, Leningrad (U.S. translation, Springer-Verlag, Berlin, 1987), 139 pp.

Callendar, G. S. 1938. The Artificial Production of Carbon Dioxide and Its Influence on Temperature, *Q. J. Roy. Meteorol. Soc. 64:* 223–240.

Callendar, G. S. 1940. Variations in the Amount of Carbon Dioxide in Different Air Currents, *Q. J. Roy. Meteorol. Soc. 66:* 395.

Callendar, G. S. 1949. Can Carbon Dioxide Influence Climate? *Weather 4:* 310–314.

CES (Committee on Earth Sciences). 1989. *Our Changing Planet: The FY 1990 Research Plan,* The U.S. Global Change Research Plan, Office of Science and Technology Policy, Washington, D.C.

Cess, R. D., G. L. Potter, J. P. Blanchet, G. J. Boer, S. J. Ghan, J. T. Kiehl, H. Le Treut, Z.-X. Li, X.-Z. Liang, J. F. B. Mitchell, J.-J. Morcrette, D. A. R. Randall, M. R. Riches, E. Roeckner, U. Schlese, A. Slingo, K. E. Taylor, W. M. Washington, R. T. Wetherald, and I. Yagai. 1989. Interpretation of Cloud-Climate Feedback as Produced by 14 Atmospheric Genereal Circulation Models, *Science 245:* 513-516.

Chamberlin, T. C. 1899. An Attempt to Frame a Working Hypothesis of the Cause of Glacial Periods on an Atmospheric Basis, *J. Geol. 7:* 545–584, 667–685, and 751–787.

Dickinson, R. E., R. J. Cicerone, 1986. Future Global Warming from Atmospheric Trace Gases, *Nature 319,* 109–115.

Emanuel, K. A. 1987. The Dependence of Hurricane Intensity on Climate, *Nature 326,* 483–485.

Frei, A., M. C. MacCracken, and M. I. Hoffert. 1988. Eustatic Sea Level and CO_2, *Northeastern Journal of Environmental Science 7(1):* 91–96.

Gilliland, R. L. 1982. Solar, Volcanic, and CO_2 Forcing of Recent Climatic Changes, *Climatic Change 4:* 111-131.

Grotch, S. L. 1988. Regional Intercomparisons of General Circulation Model Predictions and Historical Climate Data, report DOE/NBB-0084, U.S. Department of Energy, Washington, D.C.

Hansen, J., I. Fung, A. Lacis, D. Rind, S. Lebedeff, R. Ruedy, and G. Russell. 1988. Global Climate Changes as Forecast by Goddard Institute for Space Studies Three-Dimensional Model, *J. of Geophysical Research 93,* 9341–9364.

Hansen, J., D. Johnson, A. Lacis, S. Lebedeff, P. Lee, D. Rind, and G. Russell. 1981. Climate Impact of Increasing Atmospheric Carbon Dioxide, *Science 213:* 957–966.

Hansen, J., A. Lacis, D. Rind, G. Russell, P. Stone, I. Fung, R. Ruedy, and J. Lerner. 1984. Climate Sensitivity: Analysis of Feedback Mechanisms, 130–163, in J. E. Hansen and T. Takahashi (eds.), *Climate Processes and Climate Sensitivity* (Maurice Ewing Series, No. 5), American Geophysical Union, Washington, D.C., 368 pp.

Hansen, J. and S. Lebedeff. 1987. Global Trends of Measured Surface Air Temperature, *Journal of Geophysical Research 92:* 13,3345–13,372.

Hanson, K., G. A. Maul, and T. R. Karl. 1989. Are Atmospheric "Greenhouse" Effects Apparent in the Climatic Record of the Contiguous U.S. (1895–1987)?, *Geophysical Research Letters 16,* 49–52.

Hoffman, J. S., D. Keyes and J. G. Titus, 1983. *Projecting Future Sea Level Rise: Methodology, Estimates to the Year 2100, and Research Needs,* U.S. Environmental Protection Agency, Washington, D.C.

Jones, P. D., T. M. L. Wigley and P. B. Wright. 1986. Global Temperature Variations Between 1861 and 1984, *Nature 322:* 430–434.

Kerr, R. A., 1989. Hansen vs. the World on the Greenhouse Threat, *Science 244,* 1041–1043.

Lorius, C., J. Jouzel, D. Raynaud, J. Hansen, and H. LeTreut. 1989. Greenhouse Warming, Climate Sensitivity, and Ice Core Data, submitted to *Nature.*

MacCracken, M. C. 1983. Have We Detected CO_2-induced Climate Change? Problems and Prospects, *Proceedings: Carbon Dioxide Research Conference: Carbon Dioxide, Science, and Consensus,* Department of Energy report CONF-820970, Washington D.C., pp. V-3 to V-45

Manabe, S. and R. J. Stouffer. 1988. Two Stable Equilibria of a Coupled Ocean-Atmospheric Model, *J. of Climate 1,* 841–866.

Manabe, S. and R. T. Wetherald. 1967. Thermal Equilibrium of the Atmosphere with a Given Distribution of Relative Humidity, *Journal of Atmospheric Science 24:* 241–259.

Manabe, S. and R. T. Wetherald. 1987. Large Scale Changes of Soil Wetness Induced by an Increase in Atmospheric Carbon Dioxide, *Journal of Atmospheric Sciences 44:* 1211–1235.

Meier, M. 1990. Reduced Rise in Sea Level, *Nature, 343,* 115–116.

Mitchell, J. F. B., C. A. Senior, and W. J. Ingram. 1989. CO_2 and Climate: A Missing Feedback? *Nature 341,* 132–134.

NRC (National Research Council). 1975. *Understanding Climatic Change,* report of the U.S. Committee for the Global Atmospheric Research Program, National Academy of Sciences, Washington, D.C.

NRC (National Research Council). 1979. *Carbon Dioxide and Climate: A Scientific Assessment,* National Academy of Sciences, Washington, D.C.

NRC (National Research Council). 1982. *Carbon Dioxide and Climate: A Second Assessment,* National Research Council, Washington, D.C.

NRC (National Research Council). 1983. *Changing Climate,* report of the Carbon Dioxide Assessment Committee, National Academy Press.

Oort, A. H. 1983. Global Atmospheric Circulation Statistics, 1958–1973. NOAA Professional Paper 14, U.S. Government Printing Office, Washington, D.C.

Peltier, W. R. and A. M. Tushingham. 1989. Global Sea Level Rise and the Greenhouse Effect: Might They Be Connected? *Science 244,* 806–810.

PRB (Polar Research Board). 1985. *Glaciers, Ice Sheets, and Sea Level: Effects of a CO_2-Induced Climatic Change,* National Research Council, Washington, D.C.

PSAC (President's Science Advisory Council). 1965. Appendix Y4: Atmospheric Carbon Dioxide, in *Restoring the Quality of Our Environment,* Report of the Environmental Pollution Panel, The White House, Washington, D.C.

Ramanathan, V, R. D. Cess, E. F. Harrison, P. Minnis, B. R. Barkstrom, E. Ahmad and D. Hartmann. 1989. Cloud-Radiative Forcing and Climate: Results from the Earth Radiation Budget Experiment, *Science 243:* 57–63.

Revelle, R. R. 1983. Probable Future Changes in Sea Level Resulting from Increased Atmospheric Carbon Dioxide, in *Changing Climate,* Carbon Dioxide Assessment Committee, National Academy Press, Washington, D.C.

Robin, G. deQ. 1986. Changing Sea Level, in *The Greenhouse Effect, Climatic Change, and Ecosystems,* B. Bolin, B.R. Döös, J. Jäger, and R. A. Warrick (eds.), John Wiley and Sons, Chichester, pp. 323–359.

Schlesinger, M. E. and J. F. B. Mitchell. 1987. Climate Model Simulations of the Equilibrium Climatic Response to Increased Carbon Dioxide, *Review of Geophysics 25:* 760–798.

Schlesinger, M. E. and Z. C. Zhao. 1989. Seasonal Climatic Changes Induced by Doubled CO_2 as Simulated by the OSU Atmospheric GCM/Mixed-layer Ocean Model, *Journal of Climate 2:* 459–495.

Schneider, S. 1974. The Population Explosion: Can it Shake the Climate? *Ambio 3,* 150–155.

Schneider, S. H., 1989. The Greenhouse Effect: Science and Policy, *Science 243,* 771–781.

Smith, S. D. and F. W. Dobson. 1984. The Heat Budget at Ocean Weather Station Bravo. *Atmosphere-Ocean 22:* 1–22.

Somerville, R. C. J. and L. A. Remer. 1984. Cloud Optical Thickness Feedbacks in the CO_2 Climate Problem, *Journal of Geophysical Research, 89,* 9668–9672.

Sperber, K. R., S. Hameed, W. L. Gates and G. L. Potter. 1987. Southern Oscillation Simulated in a Global Climate Model, *Nature 329:* 140–142.

Washington, W. M. and G. A. Meehl. 1984. Seasonal Cycle Experiment on the Climate Sensitivity Due to a Doubling of CO_2 With an Atmospheric General Circulation Model Coupled to a Simple Mixed-Layer Ocean Model, *Journal of Geophysical Research 89:* 9475–9503.

Washington, W. M. and G. A. Meehl, 1989: Climate Sensitivity Due to Increased CO_2: Experiments with Coupled Atmosphere and Ocean General Circulation Model, *Climate Dynamics 4,* 1–38.

Washington, W. M. and C. L. Parkinson. 1986. *An Introduction to Three-Dimensional Climate Modeling,* University Science Books, Mill Valley, California.

Wilson, C. A. and J. F. B. Mitchell. 1987. A Doubled CO_2 Climate Sensitivity Experiment with a Global Climate Model Including a Simple Ocean, *J. of Geophysical Research 92,* 13,315–13,343.

Zwally, H. J., A. C. Brenner, J. A. Major, R. A. Bindschadler, and J. G. Marsh. 1989. Growth of Greenland Ice Sheet: Measurement, *Science, 246,* 1587–1589.

Chapter 5

Adams, R., M. Adams and A. Willens. 1978. *Dry Lands: Man and Plants,* The Architectural Press Ltd., London.

Armentano, T. V., R. A. Park, and C. L. Cloonan. 1986. *The Effect of Future Sea Level Rise on U.S. Coastal Wetland Areas,* Holcomb Research Institute, Butler University, Indianapolis, Indiana.

Batie, S. S. and H. H. Shugart. 1989. The Biological Consequences of Climate Changes: An Ecological and Economic Assessment, Chapter 7 in N. J. Rosenberg, Wm. Easterling, III, P. R. Crosson and J. Darmstadter, eds., *Greenhouse Warming: Abatement and Adaptation,* Resources for the Future, Washington, D.C., pp. 121–131.

Blasing, T. J. and A. M. Solomon. 1982. Response of the North American Corn Belt to Climatic Warming, Oak Ridge National Laboratory, Environmental Sciences Division, Oak Ridge, TN, Publication No. 2134, pp. 1–16.

Blasing, T. J., and A. M. Solomon. 1984. Response of the North American Corn Belt to Climatic Warming, *Progress in Biometeorology, 3:* 311-321.

Bolin, B., B. R. Döös, J. Jaeger, and R. A. Warrick, eds. 1986. *Greenhouse Effect, Climatic Change, and Ecosystems,* SCOPE 29, John Wiley and Sons, Chichester, U.K.

Box G. E. P., W. Hunter, and C. Hunter. 1987. *An Introduction to Design, Analysis, and Model Building.* John Wiley and Sons, New York, New York.

Brainard, W. 1967. Uncertainty and the Effectiveness of Policy, *American Economic Review, 57,* 411–425.

Bright, E. A., P. R. Coleman, and R. C. Durfee. 1988. *Using U.S.G.S. 1:24,000 DLG Contours to Assess the Impacts of Rising Sea Levels on a Coastal Area,* Computing and Telecommunications Division, Oak Ridge National Laboratory, Oak Ridge, Tennessee.

Brown, B. G. 1988. Climate Variability and the Colorado River Compact: Implications for Responding to Climate Change, in Glantz, M. H. (ed.) *Societal Responses to Regional Climate Change: Forecasting by Analogy,* Westview Press, Boulder, Colorado.

Bruhl, C. and P. J. Crutzen. 1988. Scenarios of Possible Changes in Atmospheric Temperature and Ozone Concentration Due to Man's Activities, Estimated with a One-dimensional Photochemical Climate Model, *Climate Dynamics 2:* 173–202.

Callaway, J. M., F. J. Cronin, J. W. Currie, and J. J. Tawil. 1982. An Analysis of Methods and Models for Assessing the Direct and Indirect Impacts of CO_2-induced Environmental Changes in the Agricultural Sector of the U.S. Economy, PNL-4384. Pacific Northwest Laboratory, Richland, Washington.

Callaway, J. M. and J. W. Currie. 1985. Water Resource Systems and Changes in Climate and Vegetation, *Characterization of Information Requirements for Studies of CO_2 Effects: Water Resources, Agriculture, Fisheries, Forests and Human Health* (M. R. White, ed.), DOE/ER-0236, U.S. Department of Energy, pp. 23-67.

Carlson, R. W. and F. A. Bazzaz. 1980. The Effects of Elevated CO_2 Concentrations on Growth, Photosynthesis, Transpiration, and Water Use Efficiency of Plants, pp. 609–623, in J. J. Singh and A. Deepak (eds.), *Environmental and Climatic Impact of Coal Utilization,* Academic Press, New York.

Chappie, M. and L. Lave. 1982. The Health Effects of Air Pollution: A Reanalysis, *Journal of Urban Economics 12(3)* :346-376.

Clark, J. S. 1988. Effect of Climate Change on Fire Regimes in Northwestern Minnesota, *Nature, 334,* 233-235.

Cohen, S. J., 1986: Impacts of CO_2-induced Climatic Change on Water Resources in the Great Lakes Basin, *Climatic Change, 8:* 135-153.

Crosson, P. R. and N. J. Rosenberg. 1989. Strategies for Agriculture, *Scientific Ameican* Special Issue on Managing Planet Earth (September 1989) pp. 128–135.

Cushman, R., J. Edmonds, W. Easterling, N. Rosenberg, M. Scott, G. Stokes, and T. Malone. 1989. *Criteria for Selecting a CO_2/Climate Change Region of Study,* Paper 89–148.7. Presented at the 82nd Annual Meeting & Exhibition, Air & Waste Management Association, Anaheim, California, June 25–30, 1989.

Davis, M. B. 1989a. Lags in Vegetation Response to Greenhouse Warming, *Climatic Change, 15:* 75–82.

Davis, M. B. 1989b. Insights from Paleoecology on Global Climate, *Bulletin of the Ecological Society of America, 70:* 222–228.

Davis, M. B. and D. B. Botkin. 1985. Sensitivity of Cool-Temperate Forests and Their Fossil Pollen Record to Rapid Temperature Change, *Quaternary Research, 23:* 327–340.

DeAngelis, D. L. and R. M. Cushman. 1989. The Potential Application of Models in Forecasting the Effects That Climate Changes Have on Fisheries, *Transactions of the American Fisheries Society* (in press).

de Ronde, John G. 1989. Past and Future Sea Level Rise in the Netherlands, A. J. Mehta and R. M. Cushman, eds. *Workshop on Sea Level Rise and Coastal Processes,* DOE/NBB-0086. US Department of Energy, Washington D.C.

Dolan, R., S. J. Trossbach, and M. K. Buckley. 1989. Pattern of Erosion Along the Atlantic Coast, *Coastal Zone '89.* American Society of Civil Engineers, pp. 17–22.

DPA Group, Inc. 1988. *CO_2-induced Climate Change in Ontario: Interdependencies and Resource Strategies,* CCD 88–09. Climate Change Digest, Environment Canada, Downsview, Ontario.

Dregne, H. E. 1983. *Desertification of Arid Lands,* Harwood Academic Publishers, New York.

Durfee, R. C., P. R. Coleman, C. D. Bonner, and E. A. Bright. 1986. *Terrain Modeling Studies to Assess Impacts of Rising Sea Level on Coastal Landmass,* Computing and Telecommunications Division, Oak Ridge National Laboratory, Oak Ridge, Tennessee.

Easterling, Wm. E. III, M. L. Parry and P. R. Crosson, 1989. Adapting Future Agriculture to Changes in Climate, Chapter 7 in N. J. Rosenberg, Wm,

Easterling III, P. R. Crosson, and J. Darmstadter (eds.), *Greenhouse Warming: Abatement and Adaptation, Resources For the Future*, Washington, D.C., pp. 91–104.

Edmonds, J. A. and J. M. Reilly. 1985. *Global Energy: Assessing the Future*, Oxford University Press, New York, New York.

Emanuel, W. R., G. G. Killough, W. M. Post, and H. H. Shugart. 1985a. Modeling Terrestrial Ecosystems in the Global Carbon Cycle with Shifts in Carbon Storage Capacity by Land-Use Change, *Ecology 65:* 970–983.

Emanuel, W. R., H. H. Shugart, and M. P. Stevenson. 1985b. Climatic Change and the Broad-Scale Distribution of Terrestrial Ecosystem Complexes, *Climatic Change* 7, 29-43.

Emanuel, K. A. 1987. The Dependence of Hurricane Intensity on Climate, *Nature 326*, 483–485.

Fair, R. 1980. Estimating the Expected Predictive Accuracy of Econometric Models, *International Economic Review*, *21*, 355–378.

Falkowski, P. G. (ed.). 1980. The Heat Budget at Ocean Weather Station Bravo. *Atmosphere-Ocean 22:* 1-22.

Flaschka, I., C. W. Stockton, and W. R. Boggess. 1987. Climatic Variation and Surface Water Resources in the Great Basin Region, *Water Resources Bulletin 23(1):* 47-57.

Frederick, K. D. and P. H. Gleick. 1989. Water Resources and Climate Change, Chapter 10 in N. J. Rosenberg, Wm. Easterling III, P. R. Crosson and J. Darmstadter (eds.), *Greenhouse Warming: Abatement and Adaptation, Resources For the Future*, Washington, D.C., pp. 133–143.

Frederick, K. D. and A. V. Kneese. 1989. Western Water Allocation Institutions and Climate Change, in P. E. Waggoner (ed.), *Climate and Water*, John Wiley, New York (in press).

Gadgil, S. et al. 1988. The Effects of Climatic Variations on Agriculture in Dry Tropical Regions of India, in Parry, M. L., T. R. Carter, and N.T. Konijin (eds.), 1988b. *The Impact of Climatic Variations on Agriculture: Volume 2. Assessments in Semiarid Regions*, Kluwer Academic Publishers, Dordrecht, The Netherlands.

Garcia, R. 1981. Drought and Man: The 1972 Case History, Volume 1, in *Nature Pleads Guilty*, Pergamon Press, New York, New York.

Gates, D. M. 1985a. Global Biospheric Response to Increasing Atmospheric Carbon Dioxide Concentration. Chapter 8 in B. R. Strain and J. D. Cure (eds.), *Direct Effects of Increasing Carbon Dioxide on Vegetation*, DOE/ER-0238, Department of Energy, Carbon Dioxide Research Division, Washington, D.C.

Gates, W. L. 1985b. The Use of General Circulation Models in the Analysis of the Ecosystem Impacts of Climatic Change, *Climatic Change*, *7*, 267-284.

Gilbert, R. O., B. A. Napier, A. M. Liebetrau, B. Sagar, and H. A. Haerer. 1990. *Statistical Aspects of Reconstructing the 131-Iodine Dose to the Thyroid of Individuals Living Near the Hanford Site in the Mid-1940s*. PNL-SA-17384-HEDR, Pacific Northwest Laboratory, Richland, Washington.

Giorgi, P., et al. 1989. Modeling the Climate of the Western U.S. with a Limited Area Model Coupled to a General Circulation Model, American Meteorological Society 6th Conference on Applied Climatology, American Meteorology Society, Boston, Massachusetts.

Glantz, M. H. (ed.) 1988. *Societal Responses to Regional Climate Change: Forecasting by Analogy*, Westview Press, Boulder, Colorado.

Glantz, M. R. Katz, and Maria Krenz (eds.) 1987. *The Social Impacts Associated With the 1982-83 Worldwide Climate Anomalies*, National Center for Atmospheric Research, Boulder, Colorado.

Gleick, P. H. 1987a. Regional Consequences of Increases in Atmospheric CO_2 and Other Trace Gases, *Climatic Change 10:* 137–161.

Gleick, P.H., 1987b. The Development and Testing of a Water Balance Model for Climate Impact Assessment: Modeling the Sacramento Basin, *Water Resources Research*, *23*, 1049-1061.

Glynn, P. and A. Manne. 1988. On the Valuation of Payoffs From a Geometric Random Walk on Oil Prices. *Pacific and Asian Journal of Energy*.

Gornitz, V. and P. Kanciruk. 1989. Assessment of Global Coastal Hazards From Sea Level Rise, *Coastal Zone '85*, American Society of Civil Engineers, pp. 1345–1359.

Graham, R. W. 1988. The Role of Climatic Change in the Design of Biological Reserves: The Paleoecological Perspective for Conservation Biology, *Conservation Biology 2(4):* 391–394.

Grigg, R. W. and D. Epp. 1989. Critical Depth For the Survival of Coral Islands: Effects on the Hawaiian Archipelago, *Science 243:* 638–641.

Grotch, S. L. 1988. Regional Intercomparisons of General Circulation Model Predictions and Historical Climate Data, report DOE/NBB-0084, U.S. Department of Energy, Washington, D.C.

Haile, D. G. 1988. *Computer Simulation of the Effects of Changes in Weather Patterns on Vector-Borne Disease Transmission,* report prepared for U.S. EPA, Office of Policy, Planning, and Evaluation. U.S. EPA Project No. DW12932662-01-1.

Hekstra, Gjerrit P. 1989. Sea Level Rise: Regional Consequences and Responses, in Rosenberg et al. (eds.), *Greenhouse Warming: Abatement and Adaptation,* Resources for the Future, Washington, D.C.

Henderson-Sellers, A. and K. McGuffie. 1986. The Threat From Melting Ice Caps, *New Scientist 110:* 24-25.

Holland, G. J., J. L. McBride, and N. Nicholls. 1987. Australian Region Tropical Cyclones and the Greenhouse Effect, pp. 438–455 in *Greenhouse,* E. J. Brill.

Hull, C. H. J. and J. G. Titus. 1986. *Greenhouse Effect, Sea Level Rise, and Salinity in the Delaware Estuary,* EPA 230-05-86-010. U.S. Environmental Protection Agency, Washington, D.C.

Hunsaker, C. T., R. L. Graham, G. W. Suter, R. V. O'Neill, B. L. Jackson, and L. W. Barnthouse, 1989. *Regional Ecological Risk Assessment: Theory and Demonstration,* ORNL/TM-11128. Oak Ridge National Laboratory, Oak Ridge, Tennessee.

Idso, S. B. and A. J. Brazel. 1984. Rising Atmospheric Carbon Dioxide Concentrations May Increase Streamflow, *Nature 312:* 51-53.

Jaeger, J. 1983. *Climate and Energy Systems: A Review of Their Interactions,* John Wiley & Sons, New York.

Kalkstein, L. S. 1988. *The Impact of CO_2 and Trace Gas-induced Climate Change Upon Human Mortality,* EPA, Office of Policy, Planning, and Evaluation. U.S. EPA Project No. CR81430101.

Kates, R. W., J. H. Ausubel, and M. Berberian (eds.). 1985. *SCOPE 27. Climate Impact Assessment, Studies of the Interaction of Climate and Society,* John Wiley and Sons, New York, New York.

Klan, M. et al. 1989. *Energy Use and Conservation Trends: 1972–1986,* PNL-67-14, Pacific Northwest Laboratory, Richland, Washington.

Kyper, T. N. and R. M. Sorensen. 1985. The Impact of Selected Sea Level Rise Scenarios On the Beach and Coastal Structures at Sea Bright, N. J., *Coastal Zone '85,* American Society of Civil Engineers, pp. 2645–2661.

Lamb, H. H. 1982. *Climate History and the Modern World,* Metheun, London, England.

Lemon, E. R., ed. 1983. *CO_2 and Plants, The Response of Plants to Rising Levels of Atmospheric Carbon Dioxide,* AAAS Selected Symposium 84, Westview Press, Inc., Boulder, Colorado.

Liebetrau, A. M., M. J. Scott, and G. Yohe. 1989 (in preparation). *Dealing With an Uncertain Future I: Coping Strategies for Modeling the Uncertain Impacts of Climate Change,* Pacific Northwest Laboratory, Richland, Washington.

Likens, G. E. 1985. The Aquatic Ecosystem and Air-Land-Water Interactions, pp. 430–435 in G. E. Likens (ed.), *An Ecosystem Approach to Aquatic Ecology, Mirror Lake and Its Environment,* Springer-Verlag, New York.

Linder, K. P. and M. J. Gibbs. 1986. The Potential Effects of Climate Change on Electric Utilities: New York Case Study Results. Presented to New York State Energy Research and Development Authority, ICF Incorporated, Washington, D.C.

Martin, P. H., N. J. Rosenberg, and M. S. McKenney. 1989. Sensitivity of Evapotranspiration in a Wheat Field, a Forest and a Grassland to Changes in Climate and Direct Effects of Carbon Dioxide, *Climate Change 14:* 117–151.

Maxwell, J. B. and L. A. Barnie. 1989. Atmosphere and Climatic Change in the Arctic and Antarctic, *Ambio, 18(1),* 42–49.

McLaughlin, S. B. 1985. Effects of Air Pollution on Forests: A Critical Review, *Journal of the Air Pollution Control Association 35:* 512–534.

Mehta, A. J. and R. M. Cushman (eds.). 1989. *Workshop on Sea Level Rise and Coastal Processes,* DOE/NBB-0086. U.S. Department of Energy.

Meisner, J. D., J. S. Rosenfeld, and H.A. Regier. 1988. The Role of Groundwater in the Impact of Climate Warming on Stream Salmonines, *Fisheries 13(3):* 2–8.

Merrill, D. 1982. Overview of Integrated Data Systems: Context, Capabilities, and Status, Proceedings of the *1982 Integrated Data Users Workshop* (R. J. Olson and N. T. Millemann, eds.), CONF-8210120, Oak Ridge Natinal Laboratory, Oak Ridge, Tennessee, pp. 3–24.

Miller, W. F., P. M. Dougherty, and G. L. Switzer, 1987: Effect of Rising Carbon Dioxide and Potential Climate Change on Loblolly Pine Distribution, Growth, Survival, and Productivity, pp. 157–187 in *The Greenhouse Effects, Climate Change, and U.S. Forests,* W. E. Shands and J. S. Hoffman, (eds.), The Conservation Foundation.

Minshall, G. W., K. W. Cummins, R. C. Petersen, C. E. Cushing, D. A. Bruns, J. R. Dedell, and R. L. Vannote. 1985. Developments in Stream Ecosystem Theory, *Canadian Journal of Fisheries and Aquatic Sciences 42(5):* 1045–1055.

Minshall, G. W., R. C. Petersen, K. W. Cummins, T. L. Bott, C. E. Cushing, J. R. Sedell, and R. L. Vannote. 1983. Interbiome Comparison of Stream Ecosystem Dynamics, *Ecological Monographs 53:* 1–25.

Mooney, H. A., P. M. Vitousek, and P. A. Matson. 1987. Exchange of Materials Between Terrestrial Ecosystems and the Atmosphere, *BioScience 238:* 926–932.

NAPAP (National Acid Precipitation Assessment Program). 1989. *Plan and Schedule for NAPAP Assessment Reports,* 1989–1990. Washington, D.C.

National Research Council. 1987. *Responding to Changes in Sea Level,* Committee on Engineering Implications of Changes in Relative Mean Sea Level, Commission on Engineering and Technical Systems, National Academy Press.

National Research Council. 1988. *Toward an Understanding of Global Change,* National Academy Press, Washington, D.C.

Neilson, R. P. 1986. High-Resolution Climatic Analysis and Southwest Biogeography, *Science 232:* 27–34.

Neilson, R. P. 1987. Biotic Regionalization and Climatic Controls in Western North America, *Vegetatio 70:* 135–147.

Nordhaus, W. and G. Yohe. 1983. Future Carbon Dioxide Emission from Fossil Fuels, in *Changing Climate,* National Research Council, Washington, D.C., pp. 87–153.

Olson, R. J. 1984 *Review of Existing Environmental and Natural Resource Data Bases,* ORNL/TM-8928. Oak Ridge National Laboratory, Oak Ridge, Tennessee.

Olson, R. J., L. J. Allison, and I. L. McCollough. 1987. *ADDNET Notebook: Documentation of the Acid Deposition Data Network (ADDNET) Data Base Supporting the National Acid Precipitation Assessment Program,* ORNL/TM-10086. Oak Ridge National Laboratory, Oak Ridge, Tennessee.

Overpeck, J. T., D. Rind, and R. Goldberg. 1990. Climate-Induced Changes in Forest Disturbance and Vegetation, *Nature 343:* 51–53.

Parry, M. L., T. R. Carter, and N. T. Konijin (eds.). 1988a. The Impact of Climatic Variations on Agriculture: Volume 1. Assessments in Cool Temperate and Cold Regions, Kluwer Academic Publishers, Dordrecht, The Netherlands.

Parry, M. L., T. R. Carter, and N. T. Konijin (eds.). 1988b. The Impact of Climatic Variations on Agriculture: Volume 2. Assessments in Semiarid Regions, Kluwer Academic Publishers, Dordrecht, The Netherlands.

Pastor, J. and W. M. Post. 1988. Response of Northern Forests to CO_2-induced Climate Change, *Nature, 334,* 55–58.

Peltier, W. R., and A. M. Tushingham. 1989. Global Sea Level Rise and the Greenhouse Effect: Might They Be Connected? *Science 244:* 806-810.

Penner, J. E., P. S. Connell, D. J. Wuebbles, and C. C. Covey. 1989. Climate Change and its Interactions with Air Chemistry: Perspectives and Research Needs, in *The Potential Effects of Global Climate Change on the United States,* J. B. Smith and D. A. Tirpak (eds.), U.S., Environmental Protection Agency report EPA-230-05-89-0##, Washington, D.C.

Peters, R. L. and J. D. S. Darling. 1985. The Greenhouse Effect and Nature Reserves, *BioScience 35(1):* 707–717.

Pickett, S. T. A. and P. S. White, eds. 1985. *The Ecology of Natural Distrubance and Patch Dynamics,* Academic Press, N.Y.

Prentice, I. C. 1986. Vegetation Responses to Past Climatic Variation, *Vegetatio 67:* 131–141.

Raloff, J. 1989. Global Smog: Newest Greenhouse Projection, *Science News, 135(17):* 262–263.

Regier, H. A. (organizer). 1988. Symposium on Climate Change and Fisheries, 118th Annual Meeting of the American Fisheries Society, Toronto, Canada.

Reilly, J. M., J. A. Edmonds, R. H. Gardner, and A. L. Brenkert, 1987. Uncertainty Analysis of the IEA/ORAU CO_2 Emissions Model, *The Energy Journal, 8(3):* 1–29.

Reimus, P. W., M. J. Apted, and A. M. Liebetrau. 1989. List and Preliminary Ranking of Key Parameters for Nuclear Waste Package Performance Assessment, *in PNC SA 0865.89-001,* Battelle, Pacific Northwest Laboratories, Richland, Washington.

Riebsame, W. 1989. *Assessing the Social Implications of Climate Fluctuation: A Guide to Climate Impact Studioes,* World Climate Impacts Programme, United Nations Environmental Program, Nairobi, Kenya.

Robin, G. deQ. 1986. Changing the Sea Level, in B. Bolin et al. (eds.), *Scope 29: The Greenhouse Effect, Climatic Change, and Ecosystems,* John Wiley & Sons, New York.

Rodhe, H. 1989. Acidification in a Global Perspective, *Ambio 18:* 155–160.

Rodhe, H. and J. Grandell. 1972. On the Removal Time of Aerosol Particles from the Atmosphere by Precipitation Scavenging, *Tellus 24( ):* 442–454.

Rosenberg, N. J. 1981. The Increasing CO_2 Concentrtion in the Atmosphere and Its Implication on Agricultural Productivity, I, Effects on Photosynthesis, Transpiration and Water Use Efficiency, *Climatic Change 3:* 265–279.

Rosenberg, N. J., M. S. McKenney, and Ph. Martin. 1989. Evpotranspiration in a Greenhouse Warmed World: A Review and a Simulation, *Agricultural and Forest Meteorology* (in press).

Rosenzweig, C., 1985: Potential CO_2-induced Climate Effects on North American Wheat-producing Regions, *Climate Change, 7,* 367-389.

Rotberg, R. I. and T. K. Rabb (eds.) 1981. *Climate and History,* Princeton University Press, Princeton, New Jersey.

Sacks, J., S. B. Schiller, and W. J. Welch. 1989. Designs for Computer Experiments, *Technometrics 31(1):* 41–48.

Schlesinger, M. E. and J. F. B. Mitchell. 1985. Model Projections of the Equilibrium Climatic Response to Increased Carbon Dioxide, Chapter 4 in M. C. MacCracken and F. M. Luther (eds.), *Projecting the Climatic Effects of Increasing Carbon Dioxide,* DOE/ER-0237, Washington, D.C. Department of Energy, Carbon Dioxide Research Division.

Schneider, S. H. and R. S. Chen. 1980. Carbon Dioxide Warming and Coastal Flooding: Physical Factors and Climatic Impact, *Annual Revue of Energy 5:* 107–140.

Sedjo, R. A. and A. M. Solomon. 1989. Climate and Forests, Chapter 8 in N. J. Rosenberg, Wm. Easterling, III, P. R. Crosson and J. Darmstadter (eds.), *Greenhouse Warming: Abatement and Adaptation, Resources for the Future,* Washington, D.C., pp. 105–119.

Sibley, T. H. and R. M. Strickland. 1985. Fisheries: Some Relationships to Climate Change and Marine Environmental Factors, in M. R. White (ed.), *Characterization of Information Requirements for Studies of CO_2 Effects: Water Resources, Agriculture, Fisheries, Forests and Human Health,* DOE/ER-0236, U.S. Department of Energy, pp. 95–143.

Singh, B. 1988 *The Implications of Climate Change for Natural Resources in Quebec,* CCD 88-08, Climate Change Digest, Environment Canada, Downsville, Ontario.

Smit, B. 1989. *Climate Warming and Canada's Comparative Position in Agriculture,* CCD 89-01, Climate Change Digest, Environment Canada, Downsview, Ontario.

Smith, J. B. and D. A. Tirpak, (eds.) 1988. *The Potential Effects of Global Climate Change on the United States, Draft Report to Congress. Volume 1: National Studies,* and *Volume 2: Regional Studies,* U.S. Environmental Protection Agency, Office of Policy, Planning, and Evaluation, Office or Research and Development.

Solomon, A. M. 1986. Transient Response of Forests to CO_2-induced Climate Change: Simulation Modeling Experiments in Eastern North America, *Oecologia, 68,* 567–579.

Southward, A. J., G. T. Boalch, and L. Maddock. 1988. Fluctuations in the Herring and Pilchard Fisheries of Devon and Cornwall Linked to Change in Climate Since the 16th Century, *J. Mar. Biol. Assoc. U.K. 68:* 423-445.

Stevenson, J. C., L. G. Ward, and M. S. Kearney. 1986. Vertical Accretion in Marshes with Varying Rates of Sea Level Rise, *Estuarine Variability*, D.A. Wolfe (ed.), Academic Press, pp. 241–259.

Strain, B. R. and J. D. Cure (eds.). 1985. *Direct Effects of Increasing Carbon Dioxide on Vegetation*, U.S. Department of Energy report DOE/ER-0238, Washington, D.C.

Titus, J. G. 1988. *Greenhouse Effect, Sea Level Rise, and Coastal Wetlands*, EPA-230-05-86-013, U.S. Environmental Protection Agency, Washington, D.C.

Titus, J. G., C. Y. Kuo, M. J. Gibbs, T. B. LaRoche, M. K. Webb, and J. O. Waddell. 1987. Greenhouse Effect, Sea Level Rise, and Coastal Drainage Systems, *Journal of Water Resources Planning and Management 113:* 216-227.

U.S. Department of Energy/EIA (1989). *PC-AEO Forecasting Model for the Annual Energy Outlook 1989: Model Documentation;* DOE/EIA-MO36, Energy Information Administration, Washington, D.C. 20585.

Waggoner, P.E., 1983: Agriculture and a Climate Changed by More Carbon Dioxide, in *Changing Climate*, National Academy Press, pp. 383-418.

Webb, T., III. 1986. Is Vegetation in Equilibrium with Climate? How to Interpret Late-Quarternary Pollen Data, *Vegetatio 67:* 75–91.

White, M. R. and I. Hertz-Picciotto. 1985. Human Health: Analysis of Climate Related to Health. *Characteristics of Information Requirements for Studies of CO_2 Effects: Water Resources, Agriculture, Fisheries, Forests, and Human Health.* DOE/ER-0236. U.S. Department of Energy. Office of Energy Research, Washington, D.C.

White, M. R. (ed), 1985: *Characterization of Information Requirements for Studies of CO_2 Effects: Water Resources, Agriculture, Fisheries, Forests and Human Health*, DOE/ER-0236, U.S. Department of Energy, Washington, D.C.

Wigley, T. M. L., J. J. Ingram, and G. Farmer (eds) 1981. *Climate and History Studies in Past Climates and Their Impact on Man*, Cambridge University Press, London, England.

Wilcoxen, P. J. 1986. Coastal Erosion and Sea Level Rise: Implications for Ocean Beach and San Francisco's Westside Transport Project, *Coastal Zone Management Journal 14:* 173–191.

Wilks, D. S., 1988: Estimating the Consequences of CO_2-induced Climatic Change on North American Grain Agriculture Using General Circulation Model Information, *Climatic Change, 13,* 19–42.

Wilson, E. O. (ed.). 1988. *Biodiversity*, National Academy Press, Washington, D.C.

Wiseman, J. and J. D. Longstreth. 1988. *The Potential Impact of Climate Change on Patterns of Infectious Disease in the United States, Background Paper and Summary of a Workshop*, Report to U.S. EPA. Project No. 68-10-7289.

Woodward, F. I. 1987. *Climate and Plant Distribution*, Cambridge University Press, Cambridge, England.

World Bank. 1984. World Development Report 1984, New York, Oxford University Press.

Chapter 6

Broecker, W. S. 1985. *How to Build a Habitable Planet*, Eldigio Press, Palisades, N.Y.

Budyko, M. I. 1974. The Method of Climate Modification (in Russian). *Meteorology and Hydrology, 2,* 91–97.

Firor, J. 1988. Public Policy and the Airborne Fraction, *Climatic Change, 12,* 103–105.

Fulkerson, W. (study leader). 1989. *Energy Technology R&D. What Could Make a Difference?* ORNL-6541/Vo. 1, Oak Ridge National Laboratory, Oak Ridge, TN.

Goldemberg, J., J. B. Johansson, A. K. N. Reddy, and R. H. Williams. 1987. *Energy for a Sustainable World*, World Resources Institute. Washington. D.C.

Harvey, L., D. D. 1989. Managing Atmospheric CO_2, *Climate Change*, in press.

Lashof, D. A. and D. A. Tirpak. 1989. *Policy Options for Stabilizing Global Climate*, Draft Report to Congress, Office of Policy Planning and Evaluation, U.S. Environmental Protection Agency.

Marchetti, C. 1975. *On Geoengineering and the CO_2 Problem*, International Institute for Applied Systems Analysis, Laxenburg, Austria, 13 pp.

Ostlie, L. D. 1988. *The Whole Tree Burner: A New Technology in Power Generation*, Energy Performance Systems, Inc., Minneapolis, Minn.

Pearman, G. I. (ed.). 1989 *Greenhouse: Planning for Climate Change,* CSIRO Publications, East Melbourne, Victoria, Australia.

Pearman, G. I. and P. Hyson. 1986. Global Transport and Inter-reservoir Exchange of Carbon Dioxide with Particular Reference to Stable Isotopic Distributions, *Journal of Atmospheric Chemistry 4:* 81–124.

Ramanathan, V. 1988. The Greenhouse Theory of Climate Change: A Test by an Inadvertent Global Experiment, *Science 240:* 293–299.

Ramanathan, V., R. D. Cess, E. F. Harrison, P. Minnis, B. R. Barkstrom, E. Ahmad, and D. Hartmann. 1989. Cloud-Radiative Forcing and Climate: Results from the Earth Radiation Budget Experiment, *Science 243:* 57–63.

Ramanathan, V., L. Callis, R. Cess, J. Hansen, I. Isaksen, W. Kuhn, A. Lacis, F. Luther, J. Mahlman, R. Reck, and M. Schlesinger. 1987. Climate-Chemical Interactions and the Effects of Changing Atmospheric Trace Gases, *Rev. Geophys. 25:* 1441–1482.

Sparrow, F. T. et al. 1988. *Carbon Dioxide from Flue Gases for Enhanced Oil Recovery,* ANL/CNSV-65, NTIS, Springfield, VA.

Steinberg, M., H. C. Cheng, and F. Horn. 1984. *A Systems Study for the Removal, Recovery, and Disposal of Carbon Dioxide from Fossil Fuel Power Plants in the U.S.,* DOE/CH/00016-2, U.S. Department of Energy, Washington, D.C.

Steinberg, M. 1989a. *An Option for the Coal Industry in Utilizing Fossil Fuel Resources with Reduced CO_2 Emissions,* BNL 42228 [Rev. 5/89], Brookhaven National Laboratory, Upton, New York.

Steinberg, M. 1989b. *Biomass and Hydrocarbon Technology for Removal of Atmospheric CO_2,* BNL 43242, Brookhaven National Laboratory, Upton, New York.

U.S. Department of Energy. 1988. *1989 Gas Mileage Guide;* DOE/CE-0019/8, Washington, D.C.

Wigley, T. M. L. 1989 Possible Climate Change Due to SO_2-derived Cloud Condensation Nuclei, *Nature 339:* 365–367.

Williams, R. H. 1989. Testimony Before the Subcommittee on Foreign Relations at the House Appropriation Committee, U.S. House of Representatives, Feb. 21, 1989.

Wolsky, A. M. and C. Brooks. 1987. Recovering CO_2 from Stationary Combustors: A Bonus for Enhanced Oil Recovery and the Environment, in *Recovery and Use of Waste CO_2 in Enhanced Oil Recovery: Proceedings of a Workshop in Denver, March 19–20, 1987,* Argonne National Laboratory Report ANL/CNSV-TM-186.

Chapter 7

CES (Committee on Earth Sciences). 1989. *Our Changing Plante: The FY 1990 Research Plan,* The U.S. Global Change Research Plan, Office of Science and Technology Policy, Washington, D.C.

Watkins, J. D. 1989. Statement before the Committee on Energy and Natural Resources, United States Senate, July 26, 1989, Washington, D.C.

Appendix 1: Author Affiliations

Eugene Aronson	Sandia National Laboratories (SNL)
David Barns	Battelle Pacific Northwest Laboratory (BPNL)
Sumner Barr	Los Alamos National Laboratory (LANL)
Cary Bloyd	Argonne National Laboratory (ANL)
Dale Bruns	Idaho National Engineering Laboratory (INEL)
Robert Cushman	Oak Ridge National Laboratory (ORNL)
Roy Darwin	Battelle Pacific Northwest Laboratory (BPNL)
Donald DeAngelis	Oak Ridge National Laboratory (ORNL)
Michael Edenburn	Sandia National Laboratories (SNL)
Jae Edmonds	Battelle Pacific Northwest Laboratory (BPNL)
William Emanuel	Oak Ridge National Laboratory (ORNL)
Dennis Engi	Sandia National Laboratories (SNL)
Michael Farrell	Oak Ridge National Laboratory (ORNL)
Jeremy Hales	Battelle Pacific Northwest Laboratory (BPNL)
Edward Hillsman	Oak Ridge National Laboratory (ORNL)
Carolyn Hunsaker	Oak Ridge National Laboratory (ORNL)
Anthony King	Oak Ridge National Laboratory (ORNL)
Albert Liebetrau	Battelle Pacific Northwest Laboratory (BPNL)
Michael MacCracken	Lawerence Livermore National Laboratory (LLNL)
Bernard Manowitz	Brookhaven National Laboratory (BNL)
Gregg Marland	Oak Ridge National Laboratory (ORNL)
Sean McDonald	Battelle Pacific Northwest Laboratory (BPNL)
Joyce Penner	Lawrence Livermore National Laboratory (LLNL)
Steve Rayner	Oak Ridge National Laboratory (ORNL)
Norman Rosenberg	Resources for the Future (RFF)
Michael Scott	Battelle Pacific Northwest Laboratory (BPNL)
Meyer Steinberg	Brookhaven National Laboratory (BNL)
Walter Westman	Lawrence Berkeley Laboratory (LBL)
Donald Wuebbles	Lawrence Livermore National Laboratory (LLNL)
Gary Yohe	Wesleyan University, Sigma Xi (WU)

Auspices

This report was prepared under the auspices of the U.S. Department of Energy by a committee drawn from the DOE national laboratories and published by the Lawrence Livermore National Laboratory (UCRL-102476 Rev. 1) under contract W7405-Eng-48.

Disclaimer